Applications of CDMA in Wireless/Personal Communications

Feher/Prentice Hall Digital and Wireless Communications Series

Carne, E. Bryan. Telecommunications Primer: Signal, Building Blocks and Networks

Feher, Kamilo. Wireless Digital Communications: Modulation and Spread Spectrum Applications

Garg, Vijay, Kenneth Smolick, and Joseph Wilkes. Applications of CDMA in Wireless/Personal Communications

Garg, Vijay and Joseph Wilkes. Wireless and Personal Communications Systems

Pelton, N. Joseph. Wireless Satellite Telecommunications: The Technology, the Market & the Regulations

Ricci, Fred. Personal Communications Systems Applications

Other Books by Dr. Kamilo Feher

Advanced Digital Communications: Systems and Signal Processing Techniques

Telecommunications Measurements, Analysis and Instrumentation

Digital Communications: Satellite/Earth Station Engineering

Digital Communications: Microwave Applications

Available from CRESTONE Engineering Books, c/o G. Breed, 5910 S. University Blvd., Bldg. C-18 #360, Littleton, CO 80121, Tel. 303-770-4709, Fax 303-721-1021, or from DIGCOM, Inc., Dr. Feher and Associates, 44685 Country Club Drive, El Macero, CA 95618, Tel. 916-753-0738, Fax 916-753-1788.

Applications of CDMA in Wireless/Personal Communications

Vijay K. Garg, PhD, PE
Distinguished Member of Technical Staff
Bell Laboratories, Lucent Technologies, Inc.

Kenneth F. Smolik, PhD, PE
Distinguished Member of Technical Staff
Bell Laboratories, Lucent Technologies, Inc.

Joseph E. Wilkes, PhD, PE
Senior Research Scientist
Bell Communications Research

To join a Prentice Hall PTR
mailing list, point to:
http://www.prenhall.com/register

Prentice Hall PTR
Upper Saddle River, NJ 07458
http://www.prenhall.com

Library of Congress Cataloging-in-Publication Data

Garg, Vijay Kumar, 1938–
    Applications of CDMA in wireless/personal communications /
Vijay K. Garg, Kenneth F. Smolick, Joseph E. Wilkes.
        p. cm. — (Feher/Prentice Hall digital and wireless
communications series)
    Includes bibliographical references and index.
    ISBN 0-13-572157-1 (case)
    1. Code division multiple access. 2. Personal communication
service systems. I. Smolick, Kenneth. II. Wilkes, Joseph E.
III. Title. IV. Series.
TK5103.45.G37 1997
621.3845—dc20   96-42194
                                                            CIP

Editorial/production supervision: *BooksCraft, Inc., Indianapolis, IN*
Cover design director: *Jerry Votta*
Cover design: *Design Source*
Acquisitions editor: *Karen Gettman*
Manufacturing manager: *Alexis R. Heydt*

Published by Prentice Hall PTR
Prentice-Hall, Inc.
A Simon & Schuster Company
Upper Saddle River, NJ 07458

The publisher offers discounts on this book when ordered in bulk quantities.
For more information, contact:
    Corporate Sales Department
    Prentice Hall PTR
    One Lake Street
    Upper Saddle River, NJ 07458
    Phone: 800-382-3419   Fax: 201-236-7141
    E-mail: corpsales@prenhall.com.

Printed in the United States of America

10 9 8 7 6 5 4 3 2

ISBN: 0-13-572157-1

Prentice-Hall International (UK) Limited, *London*
Prentice-Hall of Australia Pty. Limited, *Sydney*
Prentice-Hall Canada Inc., *Toronto*
Prentice-Hall Hispanoamericana, S.A., *Mexico*
Prentice-Hall of India Private Limited, *New Delhi*
Prentice-Hall of Japan, Inc., *Tokyo*
Simon & Schuster Asia Pte. Ltd., *Singapore*
Editora Prentice-Hall do Brasil, Ltda., *Rio de Janeiro*

Table of Contents

Preface

Over the last decade, deployment of wireless communications has been significant. In the 1980s, many analog cellular networks were implemented. These networks are already reaching capacity limits in several service areas. The wireless industry anticipated these limitations at the beginning of the 1990s. Several digital technologies were introduced to increase spectral efficiency and enhance wireless communications by adding attractive and innovative features and services such as facsimile and data transmission and various call handling features. Thus, wireless communication technology has evolved from simple first-generation analog systems for business applications to second-generation digital systems with rich features and services for residential and business environments. As the end of century approaches, a new vision of ubiquitous telecommunications for individuals is beginning to emerge. This vision, known as personal communications systems (PCS), will enable the network to deliver telecommunication services (voice, data, video, and so on) without restrictions on the user's terminal, location in the world, point of access to the network, access technology, or transport method. PCS is the challenge for the future. PCS also has a second vision, the use of the 1.8-GHz band in North America. In this book, by PCS, we mean the second vision of PCS (cellular concept at a new frequency band).

There are several reasons for the transition from wireless analog to digital technology: increased traffic, which requires greater call capacity (an explosive growth of the number of wireless telephone subscribers demands that the frequency bandwidth per call should be reduced from 30 kHz, which is currently used for analog systems); speech privacy (digital technology facilitates speech to be encrypted); new services (digital technology allows voice services to be combined with other services); and greater radio link robustness (with digital technology, improved coding techniques can be used to enhance the robustness of transmission).

Europe and Japan have developed their digital mobile communications systems by using new dedicated frequency bands. In Europe, dedicated bands were necessary since each country used a different, incompatible analog standard, and the growth of the Common Market concept required common standards throughout. In North America, however, digital technologies allow coexistence with the first-generation analog technology, since there was a common nationwide standard. Thus, the North American digital technologies enhance rather than replace the existing analog technology. Several competing digital technologies are vying for predominance in the wireless market. These include the U.S. digital cellular time-division multiple access (TDMA) system, the global system of mobile communications (GSM) that also uses TDMA with a different standard, code-division multiple access (CDMA), wideband CDMA, and several other TDMA and mixed CDMA/TDMA systems. Currently, it is projected that CDMA will be the most widely deployed digital technology in the United States; however, it may be several years before the marketplace determines the dominant technology. The purpose of this book is not to assess the advantage of one digital technology over the other. Rather, its intent is to discuss the fundamental concepts of CDMA and the application of CDMA technology to both cellular and PCS systems.

In this book, we focus on concepts of CDMA for wireless applications and the underlying network needed to support these applications for voice and data communications. Our primary emphasis is on the CDMA systems standardized by the Telecommunications Industry Association (TIA) and the Alliance for Telecommunications Industry Solutions (ATIS) as standards IS-95 and IS-665. There are, of course, other CDMA systems that have been proposed by DoCoMo, Goldengate, Interdigital, Lucent Technologies (AirLoop), and others. These systems are proprietary and are not discussed here. If you are interested in any of the these proprietary systems, contact the manufacturers directly. Other digital standards also use aspects of CDMA; for example, a TDMA system proposed by Omnipoint has been approved by ATIS for trial application. This TDMA system also includes a CDMA capability, but the underlying technology is TDMA. We discussed this system in our previous book, *Wireless and Personal Communications Systems,* Prentice Hall, 1996.

Several books on the market discuss the subject of CDMA. In writing this book, we decided to address the needs of the practicing engineer and the engineering manager by explaining the basis of CDMA and its application to wireless communications in both the cellular and the PCS environments. Students studying a course in telecommunications will also find this book useful as they prepare for a career in the wireless industry. We have incorporated sufficient mathematics so that you can understand the principles of CDMA, and yet we do not attempt to overwhelm you with mathematics. This book can be used by practicing telecommunication engineers involved in the design of cellular/PCS systems and by senior/graduate students in Electrical, Telecommunication, or Computer Engineering curricula. We do assume that you have a basic understanding of the concepts of mobile communications; if not, our previous book will provide that understanding. With selective reading of the chapters, telecommunications managers who are engaged in managing CDMA systems and who have little or no technical background in wireless technologies can gain an understanding of the systems they are managing.

This book covers several aspects of CDMA technology. In chapter 1, we explore the growth in the wireless communications and present market trends. We develop the market and technical needs for digital technologies and discuss their merits when compared to analog technology. We briefly describe the digital technologies used for cellular technology. In chapter 2, we describe different spread spectrum (SS) systems and then focus on the direct sequence spread spectrum (DSSS) technology that is specified in the TIA IS-95 and TIA-T1P1 J-STD-008 CDMA systems. We develop necessary relationships to evaluate the performance of a DSSS system with binary phase-shift keying (BPSK) and quadrature phase-shift keying (QPSK) modulation and provide a relationship to calculate the performance of a CDMA system. We conclude the chapter by discussing the main features of a CDMA system.

In chapter 3, we provide a survey of CDMA standards specifying the air interface (i.e., the messaging between the base station and the mobile station). This chapter highlights the TIA IS-95A call processing model, service configuration and negotiation, and registration by the mobile station. In chapter 4, we present the TIA TR-45/46 reference model, which is a basis for the cellular and PCS standards. We discuss the mobile switching center (MSC)—base station (BS) interface. The effects upon the architecture of a CDMA system are emphasized. We conclude the chapter by discussing the basic and supplementary services that are supported by cellular and PCS standards.

In chapter 5, we introduce the concepts of the seven-layer open-system interconnect (OSI) reference model. We describe the physical layer of the CDMA system and the wideband CDMA (W-CDMA) system. We describe the network and data link layers of the two systems in chapter 6. We discuss the signaling application layer in chapter 7. We include call flows for several typical services supplied to mobile stations using CDMA and W-CDMA. Also, we examine network operations for call origination, call termination, call clearing, mobile station registration, and mobile-assisted handoffs. In chapter 8, we discuss speech-coding algorithms that have been standardized for CDMA telephony. Note that a single speech-coding algorithm has not been adopted across the various types of access technologies since an algorithm may be customized for optimization in the context of the given access technology.

We deal with the basic guidelines for engineering a CDMA system in chapter 9. This chapter discusses several topics that are germaine to the engineering of a CDMA system. These topics include indoor and outdoor propagation models, link budgets, transitioning from analog to CDMA operation, facilities engineering, radio link capacity, and border cells located at a boundary between two service providers. In chapter 10, we concentrate on wireless data systems, including the wide area wireless data system and the high-speed Wireless Local Area Network (WLAN). We discuss the standards activities for wireless data and outline the access methods and error control schemes. We also include data services standards for wideband systems and present highlights of the TIA IS-99, TIA IS-636, and the TIA IS-657 standards. In chapter 11, we focus on the management goals for PCS networks and present the requirements for PCS network management. We discuss the important aspects of the Telecommunications Management Network (TMN) architecture, which can be applied to the management of a PCS network. We conclude the chapter by presenting requirements, as defined by TMN, for five management functions: accounting management, security management, configuration management, fault management, and performance management.

As we previously noted in this preface, the wireless industry deploys various analog and digital technologies. However, mobile subscribers expect seamless operation as they traverse different cellular/PCS systems. We examine the issues of coexistence of CDMA systems with other digital and analog systems in chapter 12. We also describe the associated work on wireless intelligent networks. The wireless industry is seeking means for improving and reducing costs so that the mobile subscriber can experience better service at a reduced price. In chapter 13, we examine several approaches that address these goals. First, service providers are seeking ways to reduce administrative costs by streamlining the service activation procedures for new mobile subscribers. We discuss over-the-air service provisioning (OTASP), which supports this objective. Second, advances in digital technology will make it possible to improve the quality of speech coding at a given data rate. We present a brief discussion of the enhanced variable rate codec (EVRC), which provides better performance than the current standardized 8-kbps speech coder. Third, the wireless industry is seeking improvements to transmission schemes used for the air interface. Resulting improvements will increase the capacity of a radio channel. For mobile subscribers, this increased capacity translates to better service at a lower price. We conclude this chapter with a discussion of three separate approaches for addressing this objective: interference cancellation, multiple beam adaptive antenna arrays, and improvements of the handoff algorithm.

We suggest material in chapters 1, 3–8, 11, and 12 for telecommunication managers. The practicing telecommunication engineer should study the entire book in order to become proficient in the CDMA technology. If this book is used for students with a general background in electromagnetic field theory and digital systems, we suggest using the material in chapters 1–12 for a one-semester course in CDMA technology.

Figures 3.1–3.4, 4.1, 4.2, 8.3, 8.4, 10.3, 10.5, 10.6, 10.8, 10.9, and 13.1 and Tables 3.1–3.5, 4.1–4.5, and 5.2–5.8 are copyrighted by the TIA and are used with permission. (To purchase the complete text of any TIA document, call Global Engineering Documents at 1-800-854-7179 or fax to 303-397-2740.) Table 9.26 is copyrighted by QUALCOMM and is used with permission. The material in chapter 9 is adapted from a Lucent Technologies Technical Education Center course and is used with permission. Some figures and tables have been adapted from our previous text and are used with permission from AT&T.

We acknowledge the many helpful suggestions we received from our reviewers, some known and some anonymous. Robert Buus, Qi Bi, Bruce McNair, Nitan Shan, and E. Lee Sneed provided useful input to help us clearly explain the CDMA concepts without overwhelming you with mathematics. We acknowledge the help of Andrew Smolick in preparing some of the tables in Chapter 9. We give special thanks to Reed Fisher, of Oki Telecom, for his many helpful discussions on the practical and theoretical aspects of CDMA and W-CDMA. One of the authors (JW) is privileged to have known Reed for 25 years as a friend and colleague and describes him as "the person I go to when I have a radio question."

Finally, we acknowledge the assistance of Karen Gettman of Prentice-Hall, Inc. and the staff of BooksCraft, Inc., received during the production of this book.

Vijay Garg
Ken Smolik
Joe Wilkes
August 1996

We dedicate this book to our families.

Vijay Garg
Ken Smolick
Joe Wilkes

Introduction to Code-Division Multiple Access Technology

1.1 INTRODUCTION

Wireless communications have shown a profound effect on our day-to-day lives. In less than 10 years, cellular telephones have attracted about a hundred million subscribers in the United States, Europe, and Asia. This dramatic development is just the start of the forthcoming revolution in telecommunication services referred to as the communication super highway. By the end of this century, telecommunication devices will be associated with the person rather than associated with a home, office, or car. To meet the unprecedented demand for a new mode of telecommunications, several digital technologies emerged at the beginning of the 1990s. All these technologies are seen as stopgaps to provide solutions to specific problems while the world waits for the technology to meet the needs of the 21st century. The vision of the third-generation personal communications network (PCN) is intended to link systems that move information among people anytime, anywhere.

In this chapter, we explore the growth in wireless communications and present market trends. We then point out the need for digital technologies and discuss their advantages over the analog technology. Next, we briefly describe the digital technologies used for cellular telephony and conclude the chapter by providing a summary of the IS-95 CDMA system.

1.2 TRENDS IN WIRELESS COMMUNICATIONS

In the 1980s, many analog cellular networks in the world reached their capacity limits. At the beginning of the 1990s, several digital technolo-

1

gies were introduced to overcome these limitations and to enhance analog cellular networks by adding several new services and features such as facsimile and data transmission and various call-handling features. As the end of 20th century approaches, a vision of the ubiquitous telecommunication services known as the PCS has developed.

There were about 36 million cellular subscribers in the U.S. in 1996. The industrialized countries in the Far East will experience a very high growth rate. It is expected that cellular subscribers in the world will grow at about 20 percent per year. Most of this growth will be handled by digital cellular networks since the analog networks have already reached or will reach their capacity limits.

In the late 1990s, PCS will grow to provide wireless access to the world telephone networks at cost-effective rates for the mass market. PCS will evolve to use a single personal number usable on the wireline and the wireless networks to provide individualized services to the subscriber anywhere, anytime. The services will be delivered using a combination of standards, networks, and products that meet a wide range of user requirements at a reasonable price.

The world is on the edge of a wireless revolution. By the year 2000, there are expected to be about 40 million new PCS users worldwide. Added to the existing cellular users, we anticipate a total of about 200 million wireless telephones. Thus, the wireless communications business presents the most exciting opportunity in telecommunications today and provides these opportunities for both service providers and equipment manufacturers.

1.3 MARKET TRENDS FOR WIRELESS COMMUNICATIONS

The wireless communications market may be a $60 billion industry at the turn of century. Cellular companies continue to see about a 20 percent growth rate in the number of subscribers. Previous predictions for wireless usage have been exceedingly low. Wireless communications is growing in many different markets. For example, in Europe and North America, we are seeing cellular and PCS as an adjunct to wireline services. At the same time, the developing countries are installing first-class telecommunications services for their citizens, with the networks interconnected to the worldwide phone network. PCS offers the opportunity to provide these modern telecommunications services without the expense of implementing the extensive copper infrastructure. For example, India needs modern telecommunications services to support the development of software for itself and other developed countries. These two diverse needs are fueling the wireless expansion around the world.

1.4 NEEDS OF DIGITAL TECHNOLOGIES FOR WIRELESS COMMUNICATIONS

The analog cellular systems were designed 15–20 years ago when the market for wireless phone service was embryonic and not well understood. As the systems using these analog technologies grew, the need for higher capacity digital technologies was identified around the world. As the digital systems were designed, the following needs were identified:

- **Large System Capacity.** An increased number of wireless subscribers requires large system capacity. Digital systems provide much higher capacity than the analog systems because of their improved spectrum efficiency.
- **Low Operations Cost.** The analog cellular networks have operation support systems to provide operation, administration, maintenance, and provisioning (OAM&P) capabilities. Each of these systems has a different user interface, uses a different computing platform, and typically manages one type of network element. As cellular/PCS service providers move into multiple networks with a mixed vendors' environment, they can no longer afford a different management system for each network element. A centralized management system is needed. The digital technology with the application of centralized management approaches lowers OAM&P cost.
- **Revenue Growth.** Digital technology facilitates implementation of new services such as data transmission, facsimile, or several supplementary services that generate additional revenues. Digital technology also can be easily adapted to provide security mechanisms.
- **User Terminal.** In general, digital terminals are smaller and lighter than analog terminals, which makes it easier to carry them around. Because less power is consumed, smaller batteries are required.

1.5 ELEMENTS OF A WIRELESS NETWORK

Public interest in wireless communication has grown rapidly. For the past several years, efforts in wireless communications have been focused on radio technology, where the goal is to increase spectrum efficiency and radio channel capacity. Expanded radio channel capacity is the key objective of a wireless network to meet demand for the indefinite future.

Significant challenges are also arising in the wireline networks to meet the needs of a high-density, high-mobility multimedia environment. As users become accustomed to wireless communications, they expect communications anywhere, anytime, in any medium. This expectation places an extra burden on the wireline networks to keep track of users' locations, to provide seamless communications as users move around, and to implement security procedures. The rapidly expanding wireless networks rely on Intelligent Network (IN) concepts to track users and deliver enhanced services. The switches and databases that perform these functions use Signaling System 7 (SS7) to connect mobile switching centers (MSCs), visitor location registers (VLRs), and home location registers (HLRs). Wireless networks interact with the existing wireline networks to extend information services to mobile stations (MSs). Figure 4.1 shows the elements of a mobile network based on today's cellular/PCS architecture. We discuss the architecture in detail in chapter 4.

Intersystem communications also use the SS7 network. This network has been standardized for the global system of mobile communications (GSM) in CCITT Q.1000 and for North American Digital Cellular in EIA/TIA IS-41. Both standards support communications between switches and data bases via SS7. In North America, the SS7 network employs 56 kbps, while in Europe 64 kbps is used.

1.6 DIGITAL TECHNOLOGIES

There are two basic strategies whereby a fixed spectrum resource can be allocated to different users: narrowband channelized systems and wideband systems.

Two narrowband systems are:

- **Frequency-division multiple access (FDMA)**, where each user is assigned to a different frequency. Guard bands are maintained between adjacent signals to minimize crosstalk between channels (see fig. 1.1).
- **Time-division multiple access (TDMA)**, where each user is assigned to a different time slot.

In a TDMA system, data from each user are carried in time intervals called time slots (see fig. 1.2). Several time slots make up a frame. Each time slot includes a preamble plus information bits addressed to various users, as shown in figure 1.3. The preamble provides identification and information that allows synchronization of the time slot at the intended receiver. Guard times are used between each user's information to minimize crosstalk between channels.

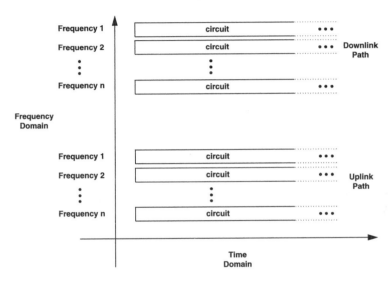

Figure 1.1 Frequency-division multiple access system.

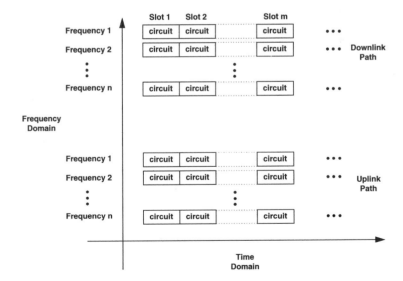

Figure 1.2 Time-division multiple access system.

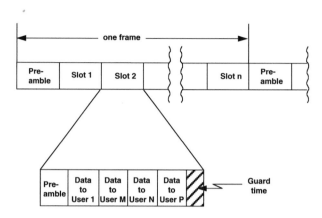

Figure 1.3 TDMA frame and time slot.

In the wideband systems, the entire system bandwidth is made available to each user and is many times larger than the bandwidth required to transmit information. Such systems are referred to as spread spectrum (SS) systems. Even though we limit our discussion to systems that use codes to select users, it is possible to design a TDMA system to have a wide bandwidth.

1.7 SPREAD SPECTRUM TECHNOLOGY

During the late 1980s and early 1990s, the rapid growth in mobile communications put a high demand on system capacity and the availability of the technology for low-cost implementation of cellular and PCS. This growth renewed the interest in commercial applications of SS mobile radios. During the same time, the Federal Communications Commission (FCC) also allowed a liberal unlicensed use of SS radios that prompted the development of a wide variety of commercial SS radio applications.

Among the many multiple-access technologies used for cellular and PCS systems, the digital SS code-division multiple access technology has been adopted as a standard in North America. The standard was developed as an alternative to the IS-54 TDMA system for cellular. The CDMA system promised improved capacity over either the analog AMPS system or the digital TDMA system.

The CDMA system, originally proposed by QUALCOMM for digital cellular phone applications, has been adopted by the Telecommunication Industry Association TR-45 committee as TIA/EIA IS-95 standard for cellular and by the Alliance for Telecommunications Industry Solutions committee T1P1 and TIA-TR46 joint standard J-STD-008 for PCS. Today, several equipment manufacturers offer CDMA systems for both cellular and PCS applications.

The major attributes of the IS-95/J-STD-008 CDMA system follow:

- **System Capacity.** The projected capacity of the CDMA system is higher than that of the existing analog system. The increased system capacity is due to improved coding gain/modulation density, voice activity, three-sector sectorization, and reuse of the same spectrum in every cell.
- **Economies.** CDMA is a cost-effective technology that requires fewer, less-expensive cells and no costly frequency reuse pattern. The average power transmitted by the CDMA mobiles averages about 6–7 mW, which is less than one tenth of the average power typically required by FM and TDMA phones. Transmitting less power means that battery life should be longer.
- **Quality of Service.** CDMA can improve the quality of service by providing robust operation in fading environments and transparent (soft) handoffs. CDMA takes advantage of multipath fading to enhance communications and voice quality. By using a RAKE receiver and other improved signal-processing techniques, each mobile station selects the three strongest multipath signals and coherently combines them to produce an enhanced signal. Thus, the fading multipath nature of the channel is used to an advantage in CDMA. In narrowband systems, fading causes a substantial degradation of signal quality.

By using a soft handoff, CDMA eliminates the ping-pong effect that occurs when a mobile nears the border between cells, and the call is rapidly switched between two cells. This effect exacerbates handoff noise, increases the load on switching equipment, and increases the chance of a dropped call. In a soft handoff, a connection is made to the new cell while maintaining the connection with the original cell. This procedure ensures a smooth transition between cells, one that is undetectable to the subscriber. In comparison, the analog and other digital systems use a break-before-make connection that increases handoff noise and the chance of a dropped call.

1.8 SUMMARY

In this chapter, we traced the growth in wireless communications and identified the need for a suitable digital technology to handle the future demand. The growth is fueled by the desire in many countries for additional telephone services that are portable/mobile and, in developing countries, as an alternative to the large investment to put copper cable into the ground.

We briefly described narrowband and wideband digital technologies and concluded the chapter by presenting the summary of the features of the IS-95/J-STD-008 SS-CDMA technology.

1.9 REFERENCES

1. Crawford, T. R., "CDMA," *Cellular Business,* June 1992, pp. 50–52.
2. Garg, V. K., and Wilkes, J. E., *Wireless and Personal Communications Systems,* Prentice Hall, Upper Saddle River, NJ, 1996.
3. Salmasi, A., "An Overview of Code-Division Multiple Access (CDMA) Applied to the Design of Personal Communications Networks," in Nanda, S., and Goodman, D. J., Ed., *Third Generation Wireless Information Network,* Kluwer Academic Publishers, Boston, 1992, pp. 277–98.
4. Singh, Y. P., "Mobile Telephony-State of Art and Future Trends in the World," *Electronic Information & Planning,* January 1995, pp. 171–75.

Spread Spectrum Systems

2.1 INTRODUCTION

In this chapter, we briefly describe the different types of spread spectrum (SS) systems that are used and then focus on the direct sequence spread spectrum (DSSS) technique that is used in code-division multiple access systems. We develop the necessary relationships to evaluate the performance of a DSSS system with binary phase-shift keying and quadrature phase-shift keying modulation and provide a relationship to calculate the performance of a CDMA system. We conclude the chapter by discussing the main features of the TIA IS-95A system. More details of the CDMA system will be given in chapters 3–8.

2.2 TYPES OF TECHNIQUES USED FOR SPREAD SPECTRUM (SS)

Since the late 1940s, SS techniques have been used for military applications in which clandestine operation is a major objective. SS techniques provide excellent immunity to interference, which may be the result of intentional jamming, and allow transmission to be hidden within background noise. Recently, SS systems have been adopted for civilian applications in wireless telephony systems.

There are three general approaches to implementing SS systems:

- **Direct sequence spread spectrum (DSSS)**, where a carrier is modulated by a digital code in which the code bit rate is much larger than the information signal bit rate (see fig. 2.1). These systems are also called pseudo-noise systems.

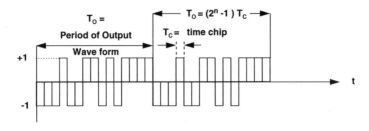

Figure 2.1 Direct sequence spread spectrum approach.

- **Frequency-hopping spread spectrum (FHSS)**, where the carrier frequency is shifted in discrete increments in a pattern generated by a code sequence (see fig. 2.2). Sometimes, the codes are chosen to avoid interference to or from other non-spread-spectrum systems. In a FHSS system, the signal frequency remains constant for a specified time duration, referred to as a time chip T_c. The FHSS system can be either a fast-hop system or a slow-hop system. In a fast-hop system, the frequency hopping occurs at a rate that is greater than the message bit rate. In a slow-hop system, the hop rate is less than the message bit rate. There is, of course, an intermediate situation in which the hop rate and message bit rate are of the same order of magnitude.

FHSS radio systems experience occasional strong bursty errors, while DSSS radio systems experience continuous but lower-level random errors. With DSSS radio systems, single errors are dis-

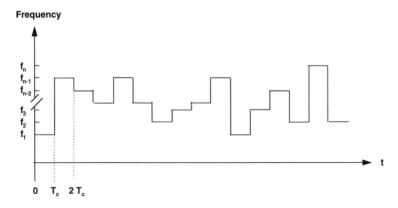

Figure 2.2 Frequency-hopping spread spectrum approach.

persed randomly over time; with FHSS radio systems, errors are distributed in clusters. Bursty errors are attributable to fading or single-frequency interference, which is time and frequency dependent. The DSSS spreads the information in both the time and frequency domains, thus providing time and frequency diversity and minimizing the effects of fading and interference.

- **Time-hopped (TH) spread spectrum**, where the transmission time is divided into intervals called frames (see fig. 2.3). Each frame is divided into time slots. During each frame, one and only one time slot is modulated with a message. All the message bits accumulated in previous frames are transmitted.

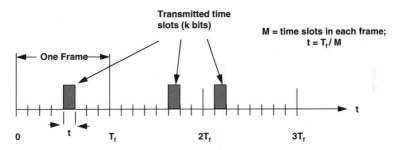

Figure 2.3 Time-hopping spread spectrum approach.

2.3 THE CONCEPT OF THE SPREAD SPECTRUM SYSTEM

The theoretical capacity of any communications channel is defined by C. E. Shannon's channel capacity formula [6]:

$$C = B_w \log_2\left[1 + \frac{S}{N}\right] \qquad (2.1)$$

where B_w = bandwidth in Hertz,
        C = channel capacity in bits per second,
        S = signal power,
        N = noise power.

Equation (2.1) gives the relationship between the theoretical ability of a channel to transmit information without errors for a given signal-to-noise (S/N) ratio and a given bandwidth on a channel. Channel capacity is increased by increasing the channel bandwidth, the transmitted power, or a combination of both.

Shannon modeled the channel at baseband. However, equation (2.1) is applicable to a radio frequency (RF) channel by assuming that the intermediate frequency (IF) filter has an ideal (flat) bandpass response with a bandwidth that is at least $2B_w$. This bound assumes that channel noise is additive white Gaussian noise (AWGN), which is often adopted in the modeling of an RF channel. This assumption is justified since the total noise is generated by random electron fluctuations. The central limit theorem provides us with the assumption that the output of an IF filter has a Gaussian distribution and is frequency independent. For most communications systems that are limited by thermal noise, this assumption is true. For interference-limited systems, this assumption is not true, and the results may be different. The Shannon equation does not provide a method to achieve the bound. Approaching the bound requires complex channel coding and modulation techniques. In many cases, achieving an implementation that provides performance near this bound is impractical due to the resulting complexity.

An analog cellular system is typically engineered to have an S/N ratio of 17 dB[1] or more. CDMA systems can be engineered to operate at much lower S/N ratios since the extra channel bandwidth can be used to achieve good performance at a very low signal-to-noise ratio.

We can rewrite equation (2.1)

$$\frac{C}{B_w} = 1.44 \log_e\left[1 + \frac{S}{N}\right] \tag{2.2}$$

since

$$\log_e\left(1 + \frac{S}{N}\right) = \frac{S}{N} - \frac{1}{2}\left(\frac{S}{N}\right)^2 + \frac{1}{3}\left(\frac{S}{N}\right)^3 - \frac{1}{4}\left(\frac{S}{N}\right)^4 + \cdots \tag{2.3}$$

we use the logarithmic expansion and assume that the S/N ratio is small (e.g., $S/N \leq 0.1$); therefore, we can neglect the higher-order terms to rewrite equation (2.2) as

$$B_w \approx \frac{C}{1.44} \times \frac{N}{S} \tag{2.4}$$

1. This ratio assumes a fading radio environment, which is typical for analog cellular systems that use frequency modulation. In the absence of fading, good FM performance is achievable at lower signal-to-noise ratios.

For any given S/N ratio, we can have a low information error rate by increasing the bandwidth used to transmit the information. As an example, if we want a system to operate on a link in which the information rate is 10 kilobits per second (kbps) and the S/N ratio is 0.01, we must use a bandwidth of

$$B_w = \frac{10 \times 10^3}{1.44 \times 0.01} = 0.69 \times 10^6 \text{ Hz or 690 kHz}$$

Information can be modulated into the SS signal by several methods. The most common method is to add the information to the spectrum-spreading code before it is used for modulating the carrier frequency (fig. 2.4). This technique applies to any SS system that uses a code sequence to determine RF bandwidth. If the signal that is being sent is analog (voice, for example), the signal must be digitized before being added to the spreading code.

One of the major advantages of an SS system is the robustness to interference. The system processing gain G_p quantifies the degree of interference rejection. The system processing gain is the ratio of RF bandwidth to the information rate and is given as

$$G_p = \frac{B_w}{R} \tag{2.5}$$

Typical processing gains for SS systems lie between 20 and 60 dB. With a SS system, the noise level is determined both by the thermal

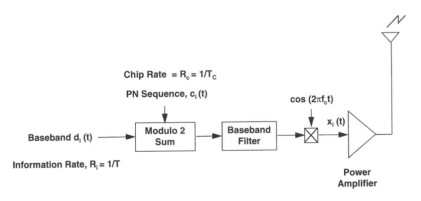

Figure 2.4 Basic DSSS system transmitter.

noise and by interference. For a given user, the interference is processed as noise. The input and output S/N ratios are related as

$$\left(\frac{S}{N}\right)_o = G_p\left(\frac{S}{N}\right)_i \tag{2.6}$$

It is instructive to relate the S/N ratio to the E_b/N_0 ratio[2] where E_b is the energy per bit and N_0 is the noise power spectral density:

$$\left(\frac{S}{N}\right)_i = \frac{E_b \times R}{N_0 \times B} = \frac{E_b}{N_0} \times \frac{1}{G_p} \tag{2.7}$$

From equation (2.6), we can express E_b/N_0 as

$$\frac{E_b}{N_0} = G_p \times \left(\frac{S}{N}\right)_i = \left(\frac{S}{N}\right)_o \tag{2.8}$$

E x a m p l e   2 . 1

Calculate the processing gain for a DSSS system that has a 10 megachips per second (Mcps) code clock rate and 4.8-kbps information rate. How much improvement in the processing gain will be achieved if the code generation rate is changed to 50 Mcps? Is there an advantage in going to a higher code generation rate with 4.8-kbps information rate?

We assume that the DSSS waveform has a voltage distribution of $(\sin x)/x$. The power distribution has a form of $[(\sin x)/x]^2$. The bandwidth of the main lobe is equal to the spreading code clock rate:

$$G_p = \frac{1.0 \times 10^7}{4.8 \times 10^3} = 2.1 \times 10^3 = 33.1 \text{ dB}$$

With 50 Mcps,

$$G_p = \frac{5 \times 10^7}{4.8 \times 10^3} = 1.04 \times 10^4 = 40.2 \text{ dB}$$

By increasing the code generation rate from 10 to 50 Mcps, we get only a 7-dB improvement in the processing gain. The effort required to increase the operating speed of a circuit by five times may be much more demanding compared to an improvement of 7 dB in the processing gain.

---

2. The noise power spectral density actually consists of both the thermal noise and interference. Unless stated explicitly, N_0 represents the thermal noise. However, common usage of this ratio assumes that N_0 includes both the thermal noise and the interference. With SS systems, interference is transformed into noise.

2.4 THE PERFORMANCE OF DIRECT SEQUENCE SPREAD SPECTRUM

2.4.1 The Direct Sequence Spread Spectrum System

The DSSS system is a wideband system in which the entire bandwidth of the system is available to each user. A system is defined to be a DSSS system if it satisfies the following requirements:

- The spreading signal has a bandwidth much larger than the minimum bandwidth required to transmit the desired information, which for a digital system is the baseband data.
- The spreading of the data is performed by means of a spreading signal, often called a code signal. The code signal is independent of the data and is of a much higher rate than the data signal.
- At the receiver, despreading is accomplished by the cross-correlation of the received spread signal with a synchronized replica of the same signal used to spread the data.

2.4.2 Coherent Binary Phase-Shift Keying

The simplest form of a DSSS communications system employs coherent binary phase-shift keying (BPSK) for both the data modulation and the spreading modulation. But, the most common form uses BPSK for the data modulation and quadrature phase-shift keying (QPSK) for the spreading modulation. We first consider the simplest case.

The encoded DSSS BPSK signal is given by

$$x(t) = c(t)s(t) = c(t)d(t)\sqrt{2S} \cos \omega_c t \qquad (2.9)$$

where $s(t) = d(t)\sqrt{2S} \cos \omega_c t$,

$d(t)$ = the baseband signal at the transmitter input
        and receiver output,

$c(t)$ = the spreading signal,

S = the signal power,

ω_c = the carrier frequency.

In equation (2.9), we represent the modulo-2 addition of $c(t)$ and $d(t)$ as a multiplication because the binary signals, 0 and 1, represent values of 1 and −1 into the modulator.

The signal $s(t)$ has a $[(\sin x)/x]^2$ spectrum of bandwidth roughly $1/T$ (where T is the periodicity at baseband), while the SS signal $x(t)$ has a similar spectrum but with a bandwidth of approximately $1/T_c$ (where T_c is the periodicity of the spreading signal). From equation (2.5), the processing gain of the system is $G_p = (B_w/R) = T/T_c$. If the interfering signal is represented by $I(t)$, then in the absence of noise (assuming that the interferer limits the system performance—in other words, that the interferer's power level exceeds the thermal noise power), the signal at the receiver is given as

$$[r(t)]^* = x(t) + I(t) \tag{2.10}$$

The receiver multiplies this by the PN waveform to obtain the signal

$$r(t) = c(t)[x(t) + I(t)] = c(t)[c(t)s(t)] + c(t)I(t) = s(t) + c(t)I(t) \tag{2.11}$$

since $c(t)^2 = 1$, $c(t)I(t)$ is the effective noise waveform due to interference. The conventional BPSK detector output is given as

$$r = d\sqrt{E_b} + n \tag{2.12}$$

where d = the data bit for the T second interval,
       E_b = the bit energy,
       n = the equivalent noise component.

The spreading-despreading operation does not affect the signal and does not affect the spectral and probability density function of the noise. For this reason, the bit error probability P_b associated with the coherent BPSK SS signal is the same as with the BPSK [13] signal and is given as

$$P_b = \frac{1}{2} \operatorname{erfc}\left(\sqrt{\frac{E_b}{N_0}}\right) \tag{2.13}$$

2.4.3 Quadrature Phase-Shift Keying

For QPSK modulation, we denote the in-phase and quadrature data waveforms as $d_c(t)$ and $d_s(t)$, respectively, and the corresponding pseudorandom noise (PN) binary waveform as $c_c(t)$ and $c_s(t)$. We can represent a QPSK signal as (see fig. 2.5)

$$x(t) = c_c(t)d_c(t)\sqrt{S} \cos \omega_c t + c_s(t)d_s(t)\sqrt{S} \sin \omega_c t \tag{2.14}$$

where each QPSK pulse is of duration $T_s = 2T$.

The in-phase output component is

$$r_c = d_c\sqrt{E_b} + n_c \tag{2.15}$$

where $n_c = \sqrt{2/T_s} \int_0^{T_c} c_c(t)I(t) \cos \omega_c t \, dt$

and the quadrature component is

$$r_s = d_s\sqrt{E_b} + n_s \tag{2.16}$$

where $n_s = \sqrt{2/T_s} \int_0^{T_c} c_s(t)I(t) \sin \omega_c t \, dt.$

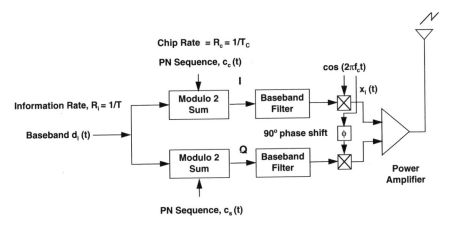

Figure 2.5 DSSS system with BPSK and QPSK transmitter.

QPSK modulation can be viewed as two independent BPSK modulations. Thus, the net data rate is doubled. We consider a special case of QPSK modulation where $c_c(t)$ and $c_s(t)$ are equal and have a value of c. The QPSK symbol energy is also the bit energy (one bit per QPSK signal).

For this case equations (2.15) and (2.16) have the form

$$r_c = d\sqrt{\frac{E_b}{2}} + n_c \tag{2.17}$$

and

$$r_s = d\sqrt{\frac{E_b}{2}} + n_s \tag{2.18}$$

where n_c and n_s are zero mean independent with conditional variances

$$\text{Var}\langle n_c | \theta \rangle = (IT_c)(\cos \theta)^2 \tag{2.19}$$

and

$$\text{Var}\langle n_s | \theta \rangle = (IT_c)(\sin \theta)^2. \tag{2.20}$$

Next we use

$$r = \frac{(r_c + r_s)}{2} = d\sqrt{\frac{E_b}{2}} + \frac{(n_c + n_s)}{2} \tag{2.21}$$

as the statistic for decision rule; then

$$\text{Var}\langle (n_c + n_s)/2 | \theta \rangle = \frac{1}{4} [IT_c(\cos \theta)^2 + IT_c(\sin \theta)^2] = \frac{IT_c}{4} \tag{2.22}$$

where θ = phase of signal

The final expression for narrowband interference $G_j(f)$ at the demodulator baseband output is given as

$$G_j(f) = \frac{IT_c}{4} = \frac{I}{4f_c} \tag{2.23}$$

For baseband systems, we define the baseband interference $I(f)$ as

$$I(f) = 2G_j(f) \quad 0 \le f \le f_c \tag{2.24}$$

The bit error probability for AWGN is given [13] as

$$P_b = \frac{1}{2} \text{erfc}\left(\sqrt{\frac{E_b}{N_0}}\right) = Q\left(\sqrt{\frac{2E_b}{N_0}}\right) \tag{2.25}$$

where $Q(u) \approx e^{-u^2/2}(\sqrt{2\pi}u)$, $u \gg 1$.

We assume that the demodulated baseband interference I is represented by AWGN.[3] For coherent PSK demodulation, we have

$$P_b = \frac{1}{2} \text{erfc}\left(\sqrt{\frac{E_b}{N_0}}\right) = \frac{1}{2} \text{erfc}\left(\sqrt{\frac{E_b}{(2I)/(4f_c)}}\right) = \frac{1}{2} \text{erfc}\left[\sqrt{2\left(\frac{S}{I}\right)\left(\frac{f_c}{f_b}\right)}\right] \tag{2.26}$$

where $I_{eff} = I/[2(f_c/f_b)]$ is referred to as the effective interference power.

---

3. This is not strictly true since the noise is known, but is sufficient for the purposes of this discussion.

The *effective interference power*, in comparison with the signal power, determines the bit error rate probability P_b of the SS system. Note that the effective interference power is reduced by the ratio of the bandwidth expansion between the baseband signal and the transmitted signal (f_c/f_b).

2.5 THE PERFORMANCE OF A CODE-DIVISION MULTIPLE ACCESS SYSTEM

A traditional narrowband system based on FDMA or TDMA is a dimension-limited system. The number of dimensions is determined by the number of non-overlapping frequencies for FDMA or by the number of time slots for TDMA. In a TDMA system, no additional users can be added once all time slots are assigned. Thus, it is not possible to increase the number of users beyond the dimension limit without causing an intolerable amount of interference to reception of a mobile station at the cell-site receiver.

Spread spectrum systems can tolerate some interference, so the introduction of each additional active radio increases the overall level of interference to the cell site receivers receiving CDMA signals from mobile station transmitters. Each mobile station introduces a unique level of interference that depends on its received power level at the cell site, its timing synchronization relative to other signals at the cell site, and its specific cross-correlation with other CDMA signals.

The number of CDMA channels in the network depends on the level of total interference that can be tolerated in the system. Thus, the CDMA system is limited by interference, and the quality of system design plays an important role in its overall capacity. A well-designed system will have a required bit error probability with a higher level of interference than a poorly designed system. Forward error-correction coding techniques improve tolerance for interference and increase the overall CDMA system capacity.

We assume that at the cell site the received signal level of each mobile user is the same and that the interference seen by each receiver is modeled as Gaussian noise. Each modulation method has a relationship that defines the bit error rate as a function of the E_b/N_0 ratio. If we know the performance of the coding methods used on the signals and tolerance of the digitized voice and the data to errors, we can define the minimum E_b/N_0 ratio for proper system operation. If we maintain operation at this minimum E_b/N_0, we can obtain the best performance of the system. The relationship between the number of mobile users M, the processing gain G_p, and the E_b/N_0 ratio are therefore given as

$$M \approx \frac{G_p}{(E_b/N_0)} \tag{2.27}$$

For a given bit error probability, the actual E_b/N_0 ratio depends on the radio system design and error-correction code. It may approach but never equal the theoretical calculations.

The best performance that can be obtained is defined by the Shannon limit[4] in AWGN. In equation (2.2), if we note that

$$\log_e\left(1 + \frac{S}{N}\right) = \frac{S}{N} - \frac{1}{2}\left(\frac{S}{N}\right)^2 + \frac{1}{3}\left(\frac{S}{N}\right)^3 - \frac{1}{4}\left(\frac{S}{N}\right)^4 + \cdots < \frac{S}{N} \tag{2.28}$$

then, from equation (2.2),

$$\frac{C}{B_w} < \frac{1}{\log_e 2}\left(\frac{S}{N}\right) \tag{2.29}$$

and

$$\frac{C}{B_w} < \frac{1}{\log_e 2}\left(\frac{E_b}{N_0}\right)\left(\frac{C}{B_w}\right). \tag{2.30}$$

Thus,

$$\frac{E_b}{N_0} \geq \log_e 2 = 0.69 = -1.59 \text{ dB} \tag{2.31}$$

provides error-free communications.

For the Shannon limit, the number of users that we can have is

$$M = \frac{G_p}{0.69} = 1.45 G_p \tag{2.32}$$

This theoretical Shannon limit shows that CDMA systems can have more users per cell than traditional narrowband systems that are limited by number of dimensions. This limit is theoretical, and in practice a wireless system is typically engineered such that E_b/N_0 = 6 dB. However, due to practical limitations on CDMA radio design, it is difficult to accommodate as many users in a single cell as given by equation (2.32). The CDMA cell capacity depends upon many factors. As seen by equation

---

4. This limit is a lower bound [6]. It is assumed that the channel coding has an infinite length to achieve this bound.

(2.32), the upper-bound theoretical capacity of an ideal noise-free CDMA channel is limited by the processing gain G_p. In an actual system, the CDMA cell capacity is much lower than the theoretical upper-bound value. The CDMA cell capacity is affected by the receiver modulation performance, power control accuracy, interference from other non-CDMA systems sharing the same frequency band, and other effects that are currently being discovered.

The CDMA transmissions in neighboring cells use the same carrier frequency and therefore cause interference that we account for by introducing a factor β. This modification reduces the number of users in a cell since the interference from users in other cells must be added to the interference generated by the other mobiles in the user's cell. The practical range for β is 0.4–0.55. The power control accuracy is represented by a factor α. The practical range for α is 0.5–0.9. We designate the reduction in the interference due to voice activity by a factor υ. The practical range for υ is 0.45–1. If directional antennas are used rather than omnidirectional antennas at the base station, the cell is sectorized with A sectors. The antennas used at the cell each radiate into a sector of $360/A$ degrees, and we have an interference improvement factor of λ. For a three-sector cell, the practical value of the improvement factor λ is 2.55. The average values for β, α, υ, and λ are taken as 0.5, 0.85, 0.6, and 2.55, respectively.

Introducing β, α, υ, and λ into equation (2.27) we get

$$M \approx \frac{G_p}{E_b/N_0} \times \frac{1}{1+\beta} \times \alpha \times \frac{1}{\upsilon} \times \lambda \qquad (2.33)$$

E x a m p l e   2 . 2

Estimate the number of mobile users that can be supported by a CDMA system that uses an RF bandwidth of 1.25 MHz to transmit data at 9.6 kbps. Assume $E_b/N_0 = 6$ dB, the interference from neighboring cells $\beta = 60\%$, the voice activity factor $\upsilon = 50\%$, and the power control accuracy factor $\alpha = 0.8$.

$$G_p = \frac{1.25 \times 10^6}{9.6 \times 10^3} = 130$$

$$\frac{E_b}{N_0} = 6 \text{ dB} = 3.98$$

$$M = \frac{130}{3.98} \times \frac{1}{1+0.6} \times \frac{1}{0.5} \times 0.8 = 32.64 \approx 33 \text{ mobile users per call}$$

The results of this example can be compared with the capacity of an analog FM system with the same frequency allocation (i.e., 41 FM channels). Typically, an analog system is engineered with a frequency reuse pattern equal to 7. With a three-sector configuration, the number of channels per sector equals $41/(7 \times 3) = 2$. This comparison suggests that a DSSS system offers a greater than tenfold improvement in the channel capacity. It is interesting to note that the processing gain of a DSSS system is directly proportional to spectrum expansion, while the processing gain of an FM system is proportional to the square of the frequency expansion.[5] This would seem to imply that the FM system should perform better than the CDMA system; yet it doesn't. There are several reasons for this CDMA performance:

- DSSS techniques take advantage of the voice activity.
- DSSS techniques use the concept of orthogonality for multiple users on a common frequency channel. This concept is applicable across different base stations and sectors.
- DSSS techniques synchronize transmission for all base stations so that soft handoffs (see chapter 7) can be implemented. This approach reduces the level of interference.

E x a m p l e   2 . 3

For the CDMA system (TIA IS-95), a chip rate[6] of 1.2288 Mcps is specified for the data rate of 9.6 kbps. E_b/N_0 is taken as 6.8 dB. Estimate the average number of subscribers that can be supported by a sector of the three-sector cell. Assume interference from neighboring cells $\beta = 50\%$, the voice activity factor $\upsilon = 60\%$, the power control accuracy factor $\alpha = 0.85$, and the improvement from sectorization $\lambda = 2.55$.

$$M \approx \frac{G_p}{E_b/N_0} \times \frac{1}{1 + \beta} \times \alpha \times \frac{1}{\upsilon} \times \lambda$$

$$G_p = \frac{(1/9.6)}{[1/(1.2288 \times 10^3)]} = 128, \quad \frac{E_b}{N_0} = 6.8 \text{ dB} = 4.7863$$

$$M = \frac{128}{4.7863} \times \frac{1}{1.5} \times \frac{1}{0.6} \times 0.85 \times 2.55 = 64.7$$

$$\text{Subscribers/Sector} = \frac{64.7}{3} = 21.57 \approx 22$$

---

5. For an FM system, the frequency expansion is specified by the deviation ratio.
6. The chip rate is the frequency of the code clock.

E X A M P L E   2 . 4

A total of 40 equal-power mobile stations are to share a frequency band through a CDMA system. Each mobile station transmits information at 9.6 kbps with a DSSS BPSK-modulated signal. Calculate the minimum chip rate of the PN code in order to maintain a bit error probability of 10^{-3}. Assume that the interference factor β from the other base stations = 60%, voice activity υ = 50%, and power control accuracy factor α = 0.8. What will be the chip rate if the probability of error is 10^{-4}?

$$P_b = Q\left(\sqrt{\frac{2E_b}{N_0}}\right) \approx \frac{e^{-E_b/N_0}}{2\sqrt{\pi(E_b/N_0)}}$$

$$\frac{e^{-E_b/N_0}}{2\sqrt{\pi(E_b/N_0)}} = 10^{-3}$$

$$\frac{E_b}{N_0} \approx 4.8 = 6.8 \text{ dB}$$

$$M = \frac{G_p}{(E_b/N_0)} \times \frac{1}{1+\beta} \times \frac{1}{\upsilon} \times \alpha$$

$$\frac{G_p}{4.8} \times \frac{1}{1.6} \times \frac{1}{0.5} \times 0.8 = 40$$

$$\therefore G_p = 192$$

$$\therefore R_c = 192 \times 9.6 \times 10^3 = 1.843 \text{ Mcps}$$

$$\text{For } P_b = 10^{-4}, \frac{E_b}{N_0} = 8.43 \text{ dB} = 6.9663$$

$$\therefore G_p = 278.5 \text{ and } R_c = 2.652 \text{ Mcps}$$

2.6 PSEUDORANDOM NOISE SEQUENCES

In CDMA systems, pseudorandom noise (PN) sequences are used to perform the following tasks:

• Spread the bandwidth of the modulated signal to the larger transmission bandwidth.
• Distinguish among the different user signals by using the same transmission bandwidth in the multiple-access scheme.

The PN sequences are not random; they are deterministic, periodic sequences. The following are the three key properties of an ideal PN sequence [2]:

• The relative frequencies of zero and one are each 1/2.
• For zeros or ones, half of all run lengths are of length 1; one quarter are of length 2; one eighth are of length 3; and so on.
• If a PN sequence is shifted by any nonzero number of elements, the resulting sequence will have an equal number of agreements and disagreements with respect to the original sequence.

PN sequences are generated by combining the outputs of feedback shift registers. A feedback shift register consists of consecutive two-state memory or storage stages and feedback logic. Binary sequences are shifted through the shift register in response to clock pulses. The contents of the stages are logically combined to produce the input to the first stage. The initial contents of the stages and feedback logic determine the successive contents of the stages. A feedback shift register and its output are called linear when the feedback logic consists entirely of modulo-2 adders.

To demonstrate the properties of PN binary sequence, we consider the linear feedback shift register (see fig. 2.6) that has a four-stage register for storage and shifting, a modulo-2 adder, and a feedback path from

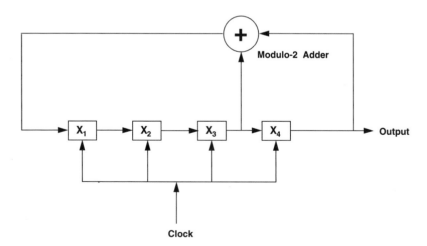

Figure 2.6 Four-stage linear feedback shift register.

adder to the input of the register. The operation of the shift register is controlled by a sequence of clock pulses. At each clock pulse, the contents of each stage in the register is shifted by one stage to the right. Also, at each clock pulse, the contents of stages X_3 and X_4 are modulo-2 added, and the result is fed back to stage X_1. The shift register sequence is defined to be the output of stage X_4. We assume that stage X_1 is initially filled with a 0, and the other remaining stages are filled with 0, 0, and 1 (i.e., the initial state of the register is 0 0 0 1). Next, we perform the shifting, adding, and feeding operations, where the results after each cycle are given in table 2.1.

Table 2.1 Results of Shifting after Each Cycle

Shift	Stage X_1	Stage X_2	Stage X_3	Stage X_4	Output Sequence
0	0	0	0	1	1
1	1	0	0	0	0
2	0	1	0	0	0
3	0	0	1	0	0
4	1	0	0	1	1
5	1	1	0	0	0
6	0	1	1	0	0
7	1	0	1	1	1
8	0	1	0	1	1
9	1	0	1	0	0
10	1	1	0	1	1
11	1	1	1	0	0
12	1	1	1	1	1
13	0	1	1	1	1
14	0	0	1	1	1
15	0	0	0	1	1
16	1	0	0	0	0

We notice that the contents of the registers repeats after $2^4 - 1 = 15$ cycles. The output sequence is given as 0 0 0 1 0 0 1 1 0 1 0 1 1 1 1 (see fig. 2.7), where the left most bit is the earliest bit. In the output sequence, the total number of zeros is 7, and the total number of ones is 8; the numbers differ by 1.

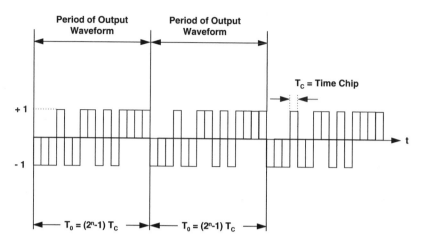

Figure 2.7 Output waveform for four-stage linear feedback shift register.

If a linear feedback shift register reached the zero state at some time, it would always remain in the zero state, and the output sequence would subsequently be all zeros. Since there are exactly $2^n - 1$ nonzero states, the period of a linear n-stage shift register output sequence cannot exceed $2^n - 1$.

The output sequences are classified as either maximal length or nonmaximal length. Maximal length sequences are the longest sequences that can be generated by a given shift register of a given length, whereas all other sequences besides maximal length sequences are nonmaximal length sequences. In the binary shift register sequence generators, the maximal length sequence is $2^n - 1$ chips, where n is the number of stages in the shift registers. Maximal length sequences have the property that, for an n-stage linear feedback shift register, the sequence repetition period in clock pulses is $T_0 = 2^n - 1$. If a linear feedback shift register generates a maximal sequence, then all its nonzero output sequences are maximal, regardless of the initial stage. A maximal sequence contains $2^{n-1} - 1$ zeros and 2^{n-1} ones per period.

2.6.1 Properties of a Maximal Length Pseudorandom Sequence

When an n-stage shift register (see fig. 2.8) is configured to generate a maximal length sequence, the sequence has the following properties:

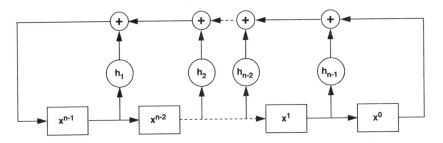

Note: $h_i = 1$ represents a closed circuit;

$h_i = 0$ represents an open circuit

Figure 2.8 n-Stage linear feedback shift register.

- The number of binary zeros differs from the number of ones by at most one chip. The number of binary ones is 2^{n-1} and the number of zeros is $2^{n-1} - 1$, where n is the number of stages in the code generator, and the code length is $2^n - 1$ chips.
- A run is defined as a sequence of a single type of binary digits. The appearance of the alternate digit in a sequence starts a new run. The length of the run is the number of digits in the run. The statistical distribution of ones and zeros is well defined and always the same. Relative positions of the runs vary from code sequence to code sequence, but the number of each run length does not.
- A modulo-2 addition of a maximal linear code with a phase-shifted replica of itself results in another replica with a phase shift different from either of the originals.
- If a period of the sequence is compared term by term with any cyclic shift itself, it is best if the number of agreements differs from the number of disagreements by not more than one count.
- If we transform the binary $(0,1)$ sequence of the shift register output to a binary $(+1, -1)$ sequence by replacing each zero by $+1$, and each 1 by -1, then the periodic correlation function of the sequence is given by

$$\theta(\tau) = \begin{bmatrix} 2^n - 1, & \tau = 0 \\ -1, & \tau \neq 0 \end{bmatrix} \qquad (2.34)$$

where τ = the shift in increments of one chip (see fig. 2.9 as
an example of $n = 3$)

n = the number of stages in the shift register.

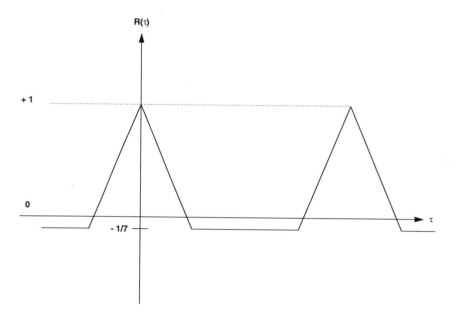

Figure 2.9 Autocorrelation of three-stage linear feedback shift register.

In the region between $\tau = 0$ and $\tau = \pm 1$, the correlation function decreases linearly from $2^n - 1$ to -1 so that the autocorrelation function for a maximal length pseudorandom sequence is triangular with a maximum value at $\tau = 0$ (see example 2.5). With this property, two or more communicators can operate independently, if their codes are phase-shifted more than one chip. For other codes sequences, the autocorrelation properties may be markedly different than the properties of the maximal length sequences.

- Every possible state of a given n-stage generator exists at some time during the generation of a complete code cycle. Each state exists for one and only one clock interval. The exception is that the all-zeros state does not normally occur and is not allowed to occur.

It has been shown [2] that there are exactly $2^{n - (p + 2)}$ runs of length p for both ones and zeros in every maximal sequence (except that there is only one run containing n ones and one containing $n - 1$ zeros; there are no runs of zeros of length n or ones of length $n - 1$). The distribution of runs for $2^4 - 1$ chip sequence is given in table 2.2.

Table 2.2 Distribution of Runs for a $2^4 - 1$ Chip Sequence

Run Length	Ones	Zeros	Number of Chips Included
1	2	2	$1 \times 2 + 1 \times 2 = 4$
2	1	1	$1 \times 2 + 1 \times 2 = 4$
3	0	1	$0 \times 3 + 1 \times 3 = 3$
4	1	0	$1 \times 4 + 0 \times 4 = 4$
Total Number of Chips			15

Whether an n-stage linear feedback shift register generates only one sequence with period $2^n - 1$ depends upon its connection vector (see fig. 2.8). Let $h(x)$ be the nth-order polynomial given by

$$h(x) = h_0 + h_1 x + h_2 x^2 + \cdots + h_n x^n \qquad (2.35)$$

We refer to $h(x)$ as the associated polynomial of the shift register with feedback coefficient $(h_0, h_1, h_2, ..., h_n)$. Here $h_0 = h_n = 1$, and other feedback coefficients take values 0 and 1. Thus, the polynomial for the four-stage linear feedback shift register, as shown in figure 2.6, is given by

$$h(x) = 1 + x^3 + x^4 \qquad (2.36)$$

When $h(x)$ is an irreducible (not factorable) primitive polynomial of degree n, then all sequences generated by $h(x)$ have a maximum period of $2^n - 1$. For an n-stage register, there are $N_p(n)$ maximal sequences that can be generated [10]. $N_p(n)$ is the number of primitive polynomials of degree n:

$$N_p(n) = \left[\frac{2^n - 1}{n} \right] \prod_{i=1}^{k} \frac{P_i - 1}{P_i} \qquad (2.37)$$

where P_i is the prime decomposition of $2^n - 1$.

Table 2.3 gives the number of maximal sequences available from register lengths 2–10 and provides an example of primitive polynomial of degree n.

2.6.2 Autocorrelation

The autocorrelation function for a signal $x(t)$ is defined as

$$R_x(\tau) = \int_{-\infty}^{\infty} x(t) x(t + \tau) dt \qquad (2.38)$$

Table 2.3 Number of Maximal Sequences Available from Register Lengths 2–10

Number of Stage n	$2^n - 1$	Prime Decomposition of $2^n - 1$	Number of n-sequence $N_p(n)$	Example of Primitive Polynomial of Degree n, $h(x)$
2	3	3	$\frac{3}{2} \cdot \frac{2}{3} = 1$	$1 + x + x^2$
3	7	7	$\frac{7}{3} \cdot \frac{6}{7} = 2$	$1 + x + x^3$
4	15	3×5	$\frac{15}{4} \cdot \frac{2}{3} \cdot \frac{4}{5} = 2$	$1 + x + x^4$
5	31	31	$\frac{31}{5} \cdot \frac{30}{31} = 6$	$1 + x^2 + x^5$
6	63	$3 \times 3 \times 7$	$\frac{63}{6} \cdot \frac{2}{3} \cdot \frac{6}{7} = 6$	$1 + x + x^6$
7	127	127	$\frac{127}{7} \cdot \frac{126}{127} = 18$	$1 + x^3 + x^7$
8	255	$3 \times 5 \times 17$	$\frac{255}{8} \cdot \frac{2}{3} \cdot \frac{4}{5} \cdot \frac{16}{17} = 16$	$1 + x^2 + x^3 + x^4 + x^8$
9	511	7×73	$\frac{511}{9} \cdot \frac{6}{7} \cdot \frac{72}{73} = 48$	$1 + x^4 + x^9$
10	1023	$3 \times 11 \times 31$	$\frac{1023}{10} \cdot \frac{2}{3} \cdot \frac{10}{11} \cdot \frac{30}{31} = 60$	$1 + x^3 + x^{10}$

Autocorrelation refers to the degree of correspondence between a sequence and a phase-shifted replica of itself. An autocorrelation plot shows the number of agreements minus disagreements for the overall length of the two sequences being compared, as the sequences assume every shift in the field of interest. If $x(t)$ is a periodic pulse waveform representing a PN sequence, we refer to each fundamental pulse as a PN sequence symbol or a chip. For such PN waveform of unit chip duration and period $T_0 = 2^n - 1$ chips, the normalized autocorrelation function is expressed as

$$R_x(\tau) = \frac{1}{T_0} \text{ [number of agreements – number of disagreements in a com-}$$

parison of one full period of sequence with a τ position cyclic shift of the sequence]

The normalized autocorrelation function $R_x(\tau)$ of a periodic waveform $x(t)$ with period T_0 is given as:

$$R_x(\tau) = \frac{1}{R_x(0)} \frac{1}{T_0} \int\limits_{-T_0/2}^{T_0/2} x(t)x(t+\tau)dt \quad \text{for } -\infty < \tau < \infty \qquad (2.39)$$

$$\text{where } R_x(0) = \frac{1}{T_0} \int\limits_{-T_0/2}^{T_0/2} x^2(t)dt$$

2.6.3 Cross-correlation

The cross-correlation function between two signals $x(t)$ and $y(t)$ is defined as the correlation between two different signals, $x(t)$ and $y(t)$, and is given as

$$R_c(\tau) = \int\limits_{-T_0/2}^{T_0/2} x(t)y(t+\tau)dt \quad \text{for } -\infty < \tau < \infty \qquad (2.40)$$

E x a m p l e   2 . 5

Consider a three-stage shift register generator, generating a seven-chip maximal linear code. The reference sequence is 1 1 1 0 0 1 0. Sketch the autocorrelation function if the chip rate is 10 Mcps.

Table 2.4 provides the sequence after each shift and shows the corresponding agreements A and disagreements D with the reference sequence.

Note that the net correlation $A - D$ is −1 for all shifts except for the zero-shift or synchronous condition. This is typical of all n sequences. In the region between zero and plus or minus one chip shift [$\tau = \pm 1/10^6$ seconds], the correlation increases linearly so that the autocorrelation function for an n sequence is triangular as shown in figure 2.9. This characteristic of auto-correlation is used to great advantage in communication systems. A channel can simultaneously support multiple users, if the corresponding codes are phase-shifted more than one chip.

2.6.4 Orthogonal Functions

Orthogonal functions are employed to improve the bandwidth efficiency of a spread spectrum system. Each mobile user uses one member of a set of orthogonal functions representing the set of symbols used for transmission. While there are many different sequences that can be used to generate an orthogonal set of functions, the Walsh and Hadamard sequences make useful sets for CDMA.

Table 2.4 Agreements and Disagreements with Reference Sequence

Shift	Sequence	Agreement, A	Disagreement, D	$A - D$
1	0 1 1 1 0 0 1	3	4	−1
2	1 0 1 1 1 0 0	3	4	−1
3	0 1 0 1 1 1 0	3	4	−1
4	0 0 1 0 1 1 1	3	4	−1
5	1 0 0 1 0 1 1	3	4	−1
6	1 1 0 0 1 0 1	3	4	−1
0	1 1 1 0 0 1 0	7	0	7

Two different methods can be used to modulate the orthogonal functions into the information stream of the CDMA signal. The orthogonal set of functions can be used as the spreading code or can be used to form modulation symbols that are orthogonal.

With orthogonal symbol modulation, the information bit stream can be divided into blocks so that each block represents a nonbinary information symbol that is associated with a particular transmitted code sequence. If there are b bits per block, one of the set of $K = 2^b$ functions is transmitted in each symbol interval. The signal at the receiver is correlated with a set of K-matched filters, each matched to the code function of one symbol. The outputs from correlators are compared and the symbol with the largest output is taken as the transmitted symbol.

If we assume a simple one-path channel with perfect power control and negligible additive noise and if we include the interference due to multipath, multiple users, and the decision process of the correlators, the E_b/N_0 ratio can be given as [8]

$$\frac{E_b}{N_0} \approx \frac{G_p}{(M - 1) + (K - 1)} \tag{2.41}$$

where M = number of mobile users,
         G_p = processing gain of the system,
         $K - 1$ = noise from the outputs of correlators other than the one corresponding to the correct symbol.

We rewrite equation (2.41) as

$$M = \frac{G_p}{E_b/N_0} - K + 2 \tag{2.42}$$

Next we introduce factors β, α, υ, and λ (see section 2.5) in equation (2.42) to get

$$M \approx \frac{G_p}{E_b/N_0} \times \frac{1}{1+\beta} \times \alpha \times \frac{1}{\upsilon} \times \lambda - K + 2 \qquad (2.43)$$

$$\eta = \frac{MR}{B_w} = \frac{M \cdot \log_2 KR_s}{B_w} = \frac{M(\log_2 K)}{G_p} \qquad (2.44)$$

where η = bandwidth efficiency,
       R_s = symbol transmission rate.

E x a m p l e   2 . 6

Calculate the bandwidth efficiency of the system using the data in Example 2.2 and assuming an orthogonal code with $K = 2$ symbols. If an orthogonal code with $K = 16$ symbols is used for the system, how many simultaneous mobile users can be supported and what is the bandwidth efficiency of the system?

For $K = 2$ symbols,

$$G_p = \frac{1.25 \times 10^6}{9.6 \times 10^3} = 130 \quad \frac{E_b}{N_0} = 6 \text{ dB} = 3.98$$

$$M \approx \frac{130}{3.98} \times \frac{1}{1+0.6} \times \frac{1}{0.5} \times 0.8 - 2 + 2 = 32.64 \approx 33 \text{ users}$$

$$\eta = \frac{M(\log_2 K)}{G_p} = \frac{33(\log_2 2)}{130} = 25.4 \text{ percent}$$

For $K = 16$ symbols,

$$M \approx \frac{130}{3.98} \times \frac{1}{1+0.6} \times \frac{1}{0.5} \times 0.8 - 16 + 2 = 18.64 \approx 19 \text{ users}$$

$$\eta = \frac{M(\log_2 K)}{G_p} = \frac{19(\log_2 16)}{130} = 58.5 \text{ percent}$$

The bandwidth efficiency of the system is improved by 33.1 percent. The disadvantage of the orthogonal signaling scheme is the complexity of the receiver design. In this example, we need 16 receiver correlators per user channel instead of only 1 required in the simplest design.

The TIA IS-95 CDMA system uses orthogonal functions[7] for the spreading code on the forward channel and orthogonal functions for the

---

7. The IS-665 wideband CDMA system uses the orthogonal codes for spreading in both directions (see chapter 5).

modulation on the reverse channel. The TIA IS-95 CDMA system uses pseudo-orthogonal function for spreads of code on the reverse. One of 64 possible modulation symbols is transmitted for each group of six code symbols. The modulation symbol is one member of the set of 64 mutually orthogonal functions. The orthogonal functions have the following characteristic:

$$\sum_{k=0}^{M-1} \phi_i(k\tau)\phi_j(k\tau) = 0 \quad i \neq j \tag{2.45}$$

where $\phi_i(k\tau)$ and $\phi_j(k\tau)$ = ith and jth orthogonal members of an
                    orthogonal set, respectively,
   M = length of the set,
   τ = symbol duration.

Walsh functions are generated by codewords rows of special square matrices called Hadamard matrices. These matrices contain one row of all zeros, and the remaining rows each have equal numbers of ones and zeros. Walsh functions can be constructed for block length $N = 2^j$, where j is an integer.

The TIA IS-95 CDMA system uses a set of 64 orthogonal functions generated by using Walsh functions. The modulated symbols are numbered from 0 through 63.

The 64×64 matrix can be generated by using the following recursive procedure:

$$H_1 = \begin{bmatrix} 0 \end{bmatrix} \quad H_2 = \begin{bmatrix} 0 & 0 \\ 0 & 1 \end{bmatrix} \tag{2.46}$$

$$H_4 = \begin{bmatrix} 0 & 0 & 0 & 0 \\ 0 & 1 & 0 & 1 \\ 0 & 0 & 1 & 1 \\ 0 & 1 & 1 & 0 \end{bmatrix} \quad H_{2n} = \begin{bmatrix} H_N & H_N \\ H_N & \overline{H_N} \end{bmatrix} \tag{2.47}$$

where N is a power of 2 and $\overline{H_N}$ is the transformation of H_N
such that 0 becomes 1 and 1 becomes 0.

The period of time needed to transmit a single modulation symbol is called a Walsh symbol interval and is equal to 1/4800 second (208.33 µs). The period of time associated with one sixty-fourth of the modulation symbol is referred to as a Walsh chip and is equal to 1/307200 second (3.255 µs). Within a Walsh symbol, Walsh chips are transmitted in the order 0, 1, 2, ..., 63.

For the forward channel, Walsh functions (figs. 2.10 and 2.11) are used to eliminate multiple access interference among users in the same cell. On downlink, all Walsh functions are synchronized in the same cell and have zero correlation between each other. Four steps are used:

1. The input user data (e.g., digital speech) are multiplied by an orthogonal Walsh function (TIA IS-95 standard uses the first 64 orthogonal Walsh functions).
2. The user data are spread by the base station pilot PN code and transmitted on the carrier.
3. At the receiver, after removing the coherent carrier, the mobile receiver multiplies the signal by the synchronized PN code (associated with the base station).

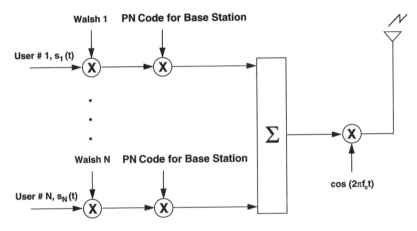

Figure 2.10 Applications of Walsh functions and offset code at the base station.

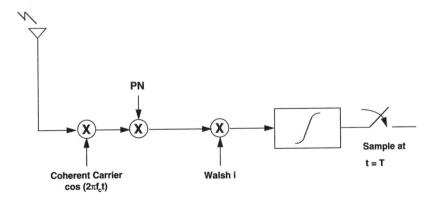

Figure 2.11 Applications of Walsh functions and offset code in the mobile station.

4. The multiplication by the synchronized Walsh function for the ith user eliminates the interference due to transmission from the base station to other users.

The Walsh functions form an ordered set of rectangular waveforms taking only two amplitudes +1 and −1. They are defined over a limited time interval T_L, which is known as the time base. If ϕ_i represents the ith Walsh function and T_L is the time base, then

$$\frac{1}{T_L} \int_0^{T_L} \phi_i(t)\phi_j(t)dt = 0 \quad \text{for } i \neq j \qquad (2.48)$$

and

$$\frac{1}{T_L} \int_0^{T_L} \phi_i^2(t)dt = 1 \quad \text{for all } i \qquad (2.49)$$

To correlate the Walsh codes at the receiver requires that the receiver be synchronized with the transmitter. In the forward direction, the base station can transmit a pilot signal to enable the receiver to recover synchronization. The designers of the IS-95 system did not believe that the pilot signal could be recovered at the base station for all mobile stations. Therefore, they used Walsh symbol modulation from the mobile station to the base station.

The designers of the IS-665 wideband CDMA system believe that, with the wider bandwidth, it will be possible to recover the pilot signal at the base station and therefore have designed a symmetric system between base station and mobile station. Chapter 5 discusses this concept in more detail.

E x a m p l e   2 . 7

We consider a case where 8 chips per bit are used to generate the Walsh functions. Specify these functions, sketch them, and show that they are orthogonal to each other.

$$H_8 = \begin{bmatrix} H_4 & H_4 \\ H_4 & \overline{H}_4 \end{bmatrix} = \begin{bmatrix} 0\ 0\ 0\ 0\ 0\ 0\ 0\ 0 \\ 0\ 1\ 0\ 1\ 0\ 1\ 0\ 1 \\ 0\ 0\ 1\ 1\ 0\ 0\ 1\ 1 \\ 0\ 1\ 1\ 0\ 0\ 1\ 1\ 0 \\ 0\ 0\ 0\ 0\ 1\ 1\ 1\ 1 \\ 0\ 1\ 0\ 1\ 1\ 0\ 1\ 0 \\ 0\ 0\ 1\ 1\ 1\ 1\ 0\ 0 \\ 0\ 1\ 1\ 0\ 1\ 0\ 0\ 1 \end{bmatrix} = \begin{bmatrix} \phi_1 \\ \phi_2 \\ \phi_3 \\ \phi_4 \\ \phi_5 \\ \phi_6 \\ \phi_7 \\ \phi_8 \end{bmatrix}$$

Figure 2.12 shows the sketches of the eight Walsh functions. We consider ϕ_2 and ϕ_4 to show orthogonality.

$$\frac{1}{T_L}\int \phi_2(t)\phi_4(t)dt = \frac{1}{T_L}[-1\times-1 + 1\times1 + 1\times-1 + 1\times(-1) + (-1)\times(-1) + 1\times1 + 1\times-1 + 1\times-1] = 0$$

and

$$\frac{1}{T_L}\int \phi_1^2(t)dt = \frac{1}{T_L}[T_L] = 1$$

Similarly, we can show that all eight Walsh functions are orthogonal to each other.

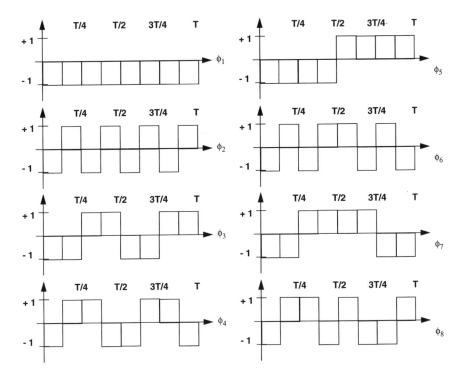

Figure 2.12 Plots of Walsh functions.

Example   2 . 8

We consider a case where eight chips per bit are used to generate the Walsh functions. Stations A, B, C, and D are assigned the chip sequence 0 1 0 1 0 1 0 1, 0 0 1 1 0 0 1 1, 0 1 1 0 0 1 1 0, and 0 0 0 0 1 1 1 1, respectively. The stations use the chip sequence to send a 1 bit and use negative chip sequences to send a 0 bit (e.g., station A uses 1 0 1 0 1 0 1 0 to send the 0 bit and so on). All chip sequences are pairwise orthogonal. This implies that the normalized correlation of any two distinct chip sequences is 0 and the normalized correlation of any chip sequence with itself is 1. We assume that all stations are synchronized in time; therefore, chip sequences begin at the same instant. When two or more stations transmit simultaneously, their bipolar signals add linearly. For example, if in one chip period three stations output +1 and one station outputs −1, the net result is +2. We consider five different cases when one or more stations transmit (see table 2.5). We want to show that the receiver recovers the bit stream of station C by computing the normalized inner products of the received sequences with the chip sequence of the station C.

Chip Sequence	Binary Values of Chip Sequence
A: 0 1 0 1 0 1 0 1	A: $(-1 +1 -1 +1 -1 +1 -1 +1)$
B: 0 0 1 1 0 0 1 1	B: $(-1 -1 +1 +1 -1 -1 +1 +1)$
C: 0 1 1 0 0 1 1 0	C: $(-1 +1 +1 -1 -1 +1 +1 -1)$
D: 0 0 0 0 1 1 1 1	D: $(-1 - 1- 1-1 +1 +1 +1 +1)$

The normalized inner products are (see table 2.5)

$$\frac{S_1 \cdot C}{8} = \frac{1+1+1+1+1+1+1+1}{8} = 1$$

$$\frac{S_2 \cdot C}{8} = \frac{2+0+0+2+0+2+2+0}{8} = 1$$

$$\frac{S_3 \cdot C}{8} = \frac{3+1+1-1+3+1+1-1}{8} = 1$$

$$\frac{S_4 \cdot C}{8} = \frac{2+0+0-2+2+0+0-2}{8} = 0$$

$$\frac{S_5 \cdot C}{8} = \frac{1-1-1-3+1-1-1-3}{8} = -1$$

Thus, the receiver recovers a bit-sequence of 1 1 1 − 0 for station C.

We assume that all the chips are synchronized in time. In a real situation, it is impossible to do so. The sender and receiver are synchronized by having the sender transmit a long enough known chip sequence that the receiver can lock onto it. All other (unsynchronized) transmissions are then seen as random noise.

Table 2.5 Five Cases

Station[a] (A B C D)	Transmitting	Received Chip Sequence
$--1-$	C	$S_1 = (-1 +1 +1 -1 -1 +1 +1 -1)$
$--1\ 1$	C + D	$S_2 = (-2\ 0\ 0 -2\ 0 +2 +2\ 0)$
$1\ 1\ 1 -$	A + B + C	$S_3 = (-3 +1 +1 +1 -3 + 1+ 1+1)$
$1\ 1 --$	A + B	$S_4 = (-2\ 0\ 0 +2 -2\ 0\ 0 +2)$
$1\ 1\ 0 -$	A + B + C	$S_5 = (-1 -1 -1 +3 -1 -1 -1 +3)$

a. Note: a dash (–) means no transmission by that station.

2.7 TIA IS-95 CDMA SYSTEM

QUALCOMM developed the CDMA radio system for digital cellular phone applications. It was optimized under existing U.S. mobile cellular system constraints. The CDMA system reuses the same frequency in all cells to increase the capacity. The *capacity* is defined as the total number of active users in a large area with many cells. This system design has been standardized by the TIA, and many equipment vendors sell CDMA equipment that meets the standard.

The IS-95 CDMA system operates in the same frequency band as the advanced mobile phone system (AMPS) using frequency-division duplex (FDD) with 25 MHz in each direction.[8] The uplink (mobile to base station) and downlink (base station to mobile) band use frequencies from 869 to 894 MHz and from 824 to 849 MHz, respectively. The mobile station supports CDMA operations on AMPS channel number 1013–1023, 1–311, 356–644, 689–694, and 739–777 inclusive. The CDMA channels are defined in terms of an RF frequency and code sequence. Sixty-four Walsh functions are used to identify the downlink channels, whereas a long PN code with different time shifts is used to identify the uplink channels. Figure 2.13 shows the CDMA channel structure. The modulation and coding features of the CDMA system are listed in table 2.6. We discuss the system in detail in chapters 3–8. There is a PCS version of the specification J-STD-008 for use in the 1800-MHz band.

---

8. The frequency spectrum for the A-System cellular service provider is split such that the spectrum is not divisible by 1.25 MHz. Thus, the A-System cellular provider cannot partition the spectrum into ten 1.25 CDMA channels. This restriction is not imposed for the B-System, however.

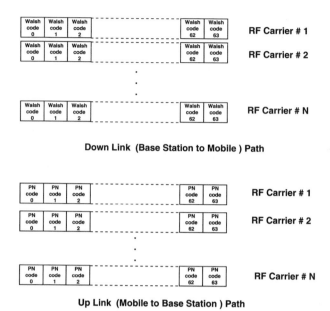

Figure 2.13 End-to-end CDMA operation at 800-MHz radio frequency.

Table 2.6 Modulation and Coding Feature of IS-95 CDMA System

Modulation	Quadrature Phase-Shift Keying
Chip rate	1.2288 Mcps
Nominal data rate	9600 bps
Filtered bandwidth	1.25 MHz
Coding	Convolution with Viterbi decoding
Interleaving	With 20-ms span

Modulation and coding details for the uplink and downlink channels differ. Pilot signals are transmitted by each cell to assist the mobile
radio in acquiring and tracking the cell site downlink signals. The strong
coding helps these radios to operate effectively at the E_b/N_0 ratio in the
5–7 dB range.

The CDMA system uses power control and voice activation to minimize mutual interference. Voice activation is provided by using a variable rate vocoder that operates at maximum rate of 8 kbps to a minimum
rate of 1 kbps, depending on the level of voice activity. With the decreased

data rate, the power control circuits reduce the transmitter power to achieve the same bit error rate. A precise power control, along with voice activation circuits, is critical to avoid excessive transmitter signal power that is responsible for contributing the overall interference in the system. Newer coding algorithms at 13 kHz will also be supported.

A time interleaver with 20-ms span is used with error-control coding to overcome rapid multipath fading and shadowing. The time span used is the same as the time frame of voice compression algorithm.

A RAKE processor is used in the CDMA radio to take advantage of a multipath delay greater than 1 µs, which is common in cellular networks in urban and suburban areas.

2.7.1 Downlink

In this section, we summarize the operation of the downlink. For more details see chapter 5. The downlink channels include 1 pilot channel, 1 synchronization (sync) channel, and 62 other channels (if multiple carriers are implemented, pilot channel and sync channels do not need to be duplicated). All the 62 channels can be used for forward traffic, but a maximum of 7 channels can be used as the paging channels. The information on each channel is modulated by an appropriate Walsh function and then modulated by a quadrature pair of PN sequences at a fixed chip rate of 1.2288 Mcps. The pilot channel is always assigned to code channel number zero. If the sync channel is present, it is given the code channel number 32. Whenever paging channels are present, they are assigned the code channel number 1 through 7 (inclusive) in sequence. The remaining code channels are used by forward traffic channels.

The sync channel always operates at a fixed data rate of 1200 bps and is convolutionally encoded to 2400 bps, repeated to 4800 bps and interleaved over the period of the pilot pseudorandom binary sequence.

The forward traffic channels are grouped into sets. Rate set 1 has four elements: 9600, 4800, 2400, and 1200 bps. Rate set 2 contains four elements: 14,400, 7200, 3600, and 1800 bps. All radio systems support rate set 1 on the forward traffic channels. Rate set 2 is optionally supported on the forward traffic channels. When a radio system supports a rate set, all four elements of the set are supported.

The speech is encoded using a variable rate vocoder to generate forward traffic channel data depending on voice activity. Since frame duration is fixed at 20 ms, the number of bits per frame varies according to the traffic rate. Since half rate convolutional encoding is used, it doubles the traffic rate to give rates from 2400 to 19, 200 symbols per second. Interleaving is performed over 20 ms. A long code of $2^{42} - 1$ ($= 4.4 \times 10^{12}$) is generated containing the user's electronic serial number (ESN) embedded in the mobile station long code mask (with voice privacy, the mobile station long code mask does not use the ESN). The scrambled data are multiplexed with power control information that steals bits from the scrambled data. The multiplexed signal remains at 19,200 bps and is changed to 1.2288 Mcps by the Walsh code W_i assigned to the ith user traffic channel. The signal is spread at 1.2288 Mcps by pilot quadrature pseudorandom binary sequence signals, and the resulting quadrature signals are then weighted. The power level of the traffic channel depends on its data transmission rate.

The paging channels provide the mobile stations with system information and instructions, in addition to acknowledging messages following access requests on the mobile stations' access channels. The paging channel data are processed in a similar manner to the traffic channel data. However, there is no variation in the power level on a per-frame basis. The 42-bit mask is used to generate the long code. The paging channel operates at a data rate of 9600 or 4800 bps.

All 64 channels are combined to give single I and Q channels. The signals are applied to quadrature modulators, and resulting signals are summed to form a QPSK signal, which is linearly amplified.

The pilot CDMA signal transmitted by a base station provides a reference for all mobile stations. It is used in the demodulation process. The pilot signal level for all base stations is 4–6 dB higher than a traffic channel with a constant value. The pilot signals are quadrature pseudorandom binary sequence signals with a period of 32,768 chips. Since the chip rate is 1.2288 Mcps, the pilot pseudorandom binary sequence corresponds to a period of 26.66 ms, which is equivalent to 75 pilot channel code repetitions every 2 seconds. The pilot signals from all base stations use the same pseudorandom binary sequence, but each base station is identified by a unique time offset of its pseudorandom binary sequence. These offsets are in increments of 64 chips providing 511 unique offsets relative to zero offset code. These large numbers of offsets ensure that unique base station identification can be obtained, even in dense microcellular environments.

A mobile station processes the pilot channel to find the strongest signal components. The processed pilot signal provides an accurate estimation of time delay, phase, and magnitude of the three multipath components. The three components are tracked in the presence of fast fading and coherently combined to improve signal quality. The chip rate on the pilot channel and on all channels is locked to precise system time, for example, by using the global positioning system (GPS). Once the mobile station identifies the strongest pilot offset by processing the multipath components from the pilot channel correlator, it examines the signal on its sync channel, which is locked to the pseudorandom binary sequence signal on the pilot channel. Since the sync channel is time aligned with its base station's pilot channel, the mobile station finds the information pertinent to this particular base station on the sync channel. The sync channel message contains time of day and long code synchronization to ensure that long code generators at the base station and mobile station are aligned and identical. The mobile station now attempts to access the paging channel and listens for system information. The mobile station enters the idle state when it has completed acquisition and synchronization. It listens to the assigned paging channel and is able to receive and initiate the calls. When informed by the paging channel that voice traffic is available on a particular channel, the mobile station recovers the speech data by applying the inverse of the spreading procedures.

2.7.2 Uplink (Reverse)

In this section, we summarize the operation of the uplink. For more details see chapter 5. The uplink channel is separated from the downlink channel by 45 MHz at cellular frequencies and 80 MHz at PCS frequencies. The uplink uses the same 32,768 chip code as is used on the downlink. The uplink channels are either access channel or reverse traffic channels. There are 62 traffic channels and up to 32 access channels (if multiple carriers are used, it is possible to assign 64 traffic channels to some of them). The access channel enables the mobile station to communicate nontraffic information (e.g., originate calls and respond to paging). The access rate is fixed at 4800 bps. All mobile stations accessing a radio system share the same frequency assignment. Each access channel is identified by a distinct access channel long-code sequence having an access number, a paging channel number associated with the access channel, and other system data. Each mobile station uses a different time shift on the PN code; therefore, the radio system can correctly decode the information from an individual mobile station. Data transmitted on the reverse channel are grouped into 20-ms frames. All data on the reverse channel are convolutionally encoded, block interleaved, and modulated by modulation symbols transmitted for each of the six code symbols. The modulation symbol is one of the 64 mutually orthogonal waveforms that are generated using Walsh functions.

The reverse traffic channel may use either 9600-, 4800-, 2400-, or 1200-bps data rates for transmission. The duty cycle for transmission varies proportionally with data rate, being 100 percent at 9600 bps to 12.5 percent at 1200 bps. An optional second rate set is also supported in the PCS version of CDMA and new versions of cellular CDMA (see chapter 5). The actual burst transmission rate is fixed at 28,800 code symbols per second. Since six code symbols are modulated as one of 64 modulation symbols for transmission, the modulation symbol transmission rate is fixed at 4800 modulation symbols per second. This results in a fixed Walsh chip rate of 307.2 kcps. The rate of spreading PN sequence is fixed at 1.2288 Mcps, so that each Walsh chip is spread by 4 PN chips. Table 2.7 provides the signal rates and their relationship for the various transmission rates on the reverse traffic channel.

Table 2.7 CDMA Reverse Traffic Channel Modulation Parameters

Parameter	9600 bps	4800 bps	2400 bps	1200 bps	Units
PN chip rate	1.2288	1.2288	1.2288	1.2288	Mcps
Code rate	1/3	1/3	1/3	1/3	bits/code symbol
Transmitting duty cycle	100	50	25	12.5	%
Code symbol rate	3 × 9600 = 28,800	28,800	28,800	28,800	sps
Modulation	6	6	6	6	code symbol/ modulation symbol
Modulation symbol rate	28,800/6 = 4800	4800	4800	4800	sps
Walsh chip rate	64 × 4800 = 307.2	307.2	307.2	307.2	kcps
Modulation symbol duration	1/4800 = 208.33	208.33	208.33	208.33	μs
PN chips/code symbol	12,288/ 288 = 42.67	42.67	42.67	42.67	PN chip/ code symbol

Table 2.7 CDMA Reverse Traffic Channel Modulation Parameters (Continued)

Parameter	9600 bps	4800 bps	2400 bps	1200 bps	Units
PN chips/modulation symbol	1,228,800/ 4800 = 256	256	256	256	PN chip/ modulation symbol
PN chips/Walsh chip	4	4	4	4	PN chips/ Walsh chip

Following the orthogonal spreading, the reverse traffic channel and access channel are spread in quadrature. Zero-offset I and Q pilot PN sequences are used for spreading. These sequences are periodic with 2^{15} (32,768 PN chips in length) chips and are based on characteristic polynomials $g_I(x)$ and $g_Q(x)$ (see equations [5.20] and [5.21]).

The maximum-length linear feedback register sequences $I(n)$ and $Q(n)$, based on these polynomials, have a period $2^{15} - 1$ and are generated by using the following recursions:

$$I(n) = I(n - 15) \oplus I(n - 8) \oplus I(n - 7) \oplus I(n - 6) \oplus I(n - 2) \quad (2.50)$$

based on $g_I(x)$ as the characteristic polynomial, and

$$q(n) = q(n - 15) \oplus q(n - 12) \oplus q(n - 11) \oplus \\ q(n - 10) \oplus q(n - 9) \oplus q(n - 5) \oplus \\ q(n - 4) \oplus q(n - 3) \quad (2.51)$$

based on $q_Q(x)$ as the characteristic polynomial, where $I(n)$ and $Q(n)$ are binary numbers (0 and 1) and the additions are modulo-2. To obtain the I and Q pilot sequences, a 0 is inserted in $I(n)$ and $Q(n)$ after 14 consecutive 0 outputs (this occurs only once in each period). Therefore, the pilot PN sequences have one run of 15 consecutive 0 outputs instead of 14. The chip rate for the pilot PN sequence is 1.2288 Mcps, and its period is 26.666 ms. There are exactly 75 repetitions in every 2 seconds. The spreading modulation is offset quadrature phase-shift keying (O-QPSK). The data spread by Q pilot PN sequence is delayed by half a chip time (406.901 ns) with respect to the data spread by I pilot PN sequence (see Chapter 5). Figure 2.14 and Table 2.8 describe the characteristics of O-QPSK.

The Wideband CDMA has some differences in the modulation methods. For more details on both systems see chapter 5.

Table 2.9 defines the signal rates and their relationship on the access channel.

Each base station transmits a pilot signal of constant power on the same frequency. The received power level of the received pilot signal

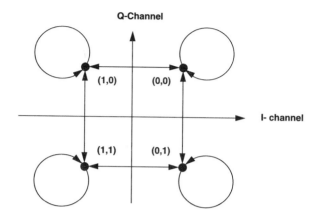

Figure 2.14 Signal constellation and phase transition of O-QPSK use on reverse CDMA channel.

Table 2.8 Reverse CDMA Channel *I* and *Q* Mapping

I	*Q*	Phase
0	0	$\pi/4$
1	0	$(3\pi)/4$
1	1	$-(3\pi)/4$
0	1	$\pi/4$

Table 2.9 CDMA Access Channel Modulation Parameters

Parameter	4800 bps	Units
PN chip rate	1.2288	Mcps
Code rate	1/3	bits/code symbol
Code symbol repetition	2	symbols/code symbol
Transmit duty cycle	100	%
Code symbol rate	28,800	sps
Modulation	6	code symbol/modulation symbol
Modulation symbol rate	4800	sps
Walsh chip rate	307.2	kcps
Modulation symbol duration	208.33	µs
PN chips/code symbol	42.67	PN chip/code symbol
PN chips/modulation symbol	256	PN chip/modulation symbol
PN chips/Walsh chip	4	PN chips/Walsh chip

enables the mobile station to estimate the path loss between the base station and the mobile station. Knowing the path loss, the mobile station adjusts its transmitted power such that the base station will receive the signal at the requisite power level. The base station measures the mobile station received power and tells the mobile station to make the necessary adjustment to its transmitter power. One command every 1.25 ms adjusts the transmitted power from the mobile station in the step of ±0.5 dB. The base station uses frame errors reported by the mobile station to increase or decrease the transmitted power.

CDMA provides a soft handoff. As the mobile station moves to the edge of its single cell, the adjacent base station assigns a modem to the call, while the current base station continues to handle the call. The call is handled by both base stations on a make-before-break basis. Handoff diversity occurs with both base stations handling the call until the mobile station moves sufficiently close to one of the base stations, which then exclusively handles the call. This handoff procedure is different than conventional break-before-make or hard handoff procedures. The soft handoff procedure will be discussed in more detail in chapter 7.

In summary, a CDMA system operates with a low E_b/N_0 ratio, exploits voice activity, and uses sectorization of cells. Each sector has 64 CDMA channels. It is a synchronized system with three receivers to provide path diversity at the mobile station and four receivers at the cell site.

2.8 SUMMARY

In this chapter, we first discussed the concept of spread spectrum systems and provided the main features of the direct sequence spread spectrum system used in the IS-95 and IS-665 systems. A key component of spread spectrum performance is the calculation of processing gain of the system, which is the relationship between the input and output signal-to-noise ratio of a spread spectrum receiver. We used the relationship to present some examples that evaluate the performance of a CDMA spread spectrum system.

We presented the Shannon equation for error-free communications and used it to determine that error-free communications is possible (with high delays) for a bit energy-to-noise density ratio, $E_b/N_0 = -1.59$ dB. Spread spectrum systems trade bandwidth for processing gain, and code division systems use a variety of orthogonal or almost orthogonal codes to allow multiple users in the same bandwidth. Thus, CDMA systems can have higher capacity than either analog or TDMA digital systems. However, because of practical constraints on CDMA systems, it is not possible to achieve the Shannon bound in system design. The upper bound of the capacity of a CDMA system is limited by the processing gain of the system. In an actual system, the capacity is lower than the theoretical upper bound. CDMA capacity is affected by receiver modulation performance, power control accuracy, interference from other cells, voice activity, cell sectorization, and the ability to maintain synchronization of the systems. Practical CDMA systems are designed for a value of $E_b/N_0 = 6$ dB.

We described the properties of ideal and practical pseudorandom noise sequences that are used in spread spectrum systems. We then described the Walsh and Hadamard functions that are used in the IS-95 and IS-665 CDMA standards for cellular and PCS. We concluded the chapter by discussing the high-level features of the IS-95 system. We will study the system in more detail in chapters 3–8.

2.9 REFERENCES

1. Bhargava, V., Haccoum, D., Matyas, R., and Nuspl, P., *Digital Communications by Satellite*, John Wiley & Sons, New York, 1981.
2. Dixon, R. C., *Spread Spectrum Systems*, Second Edition, John Wiley & Sons, New York, 1984.
3. Feher, K., *Wireless Digital Communications Modulation and Spread Spectrum Applications*, Prentice Hall PTR, Upper Saddle River, NJ, 1995.
4. Lee, W. C. Y., *Mobile Cellular Telecommunication Systems*, McGraw-Hill, New York, 1989.
5. Pahlwan, K., and Levesque, A. H., *Wireless Information Networks*, John Wiley & Sons, New York, 1995.
6. Shannon, C. E., "Communications in the Presence of Noise," *Proceedings of the IRE*, 1949, No. 37, pp. 10–21.
7. Steele, R., *Mobile Radio Communications*, IEEE Press, New York, 1992.
8. Skalar, B., *Digital Communications—Fundamental & Applications*, Prentice Hall, Englewood Cliffs, NJ, 1988.
9. Torrien, D., *Principle of Secure Communication Systems*, Artech House, Boston, 1992.
10. Viterbi, A. J., *CDMA*, Addison-Wesley Publishing Company, Reading, MA, 1995.
11. Viterbi, A. J., and Padovani R., "Implications of Mobile Cellular CDMA," *IEEE Communication Magazine*, 1992, Vol. 30, No. 12, pp. 38–41.
12. TIA/EIA IS-95, "Mobile Station–Base Station Compatibility Standard for Dual-mode Wideband Spread Spectrum Cellular System," PN-3422, 1994.
13. Garg, V. K., and Wilkes, J. E., *Wireless and Personal Communications Systems*, Prentice Hall, Upper Saddle River, NJ, 1996.

CDMA Standards

3.1 INTRODUCTION

The standardization of North American cellular operation commenced in the 1970s with the development of analog specifications. Until the late 1980s, analog technology was adequate in satisfying the needs of the market. As penetration of cellular service increased, particularly in urban areas such as Los Angeles and New York, the demand for greater call capacity increased. Thus, time-division multiple access technology was standardized. Even though TDMA offered roughly a threefold increase in the capacity, numerous service providers thought that this increase was inadequate for the future growth in service. Consequently, other technical alternatives were considered by the cellular industry. This chapter examines the standard for code-division multiple access as adopted by the Telecommunications Industry Association for cellular and later PCS systems. The chapter provides an overview of the CDMA standard; later chapters (4–8) cover the standards in more detail.

3.2 BACKGROUND

In 1992, the TIA formed the technical committee TR45.5 to study and to generate cellular standards (800 MHz) for wideband service. At that time, the existing cellular service in North America supported analog and TDMA technologies with a frequency bandwidth of 30 kHz. The TIA desired a wider bandwidth and recommended that a bandwidth greater than 1 MHz be an objective of any standard to be considered by the

TR45.5. The resulting standard became Standard TIA IS-95 (Mobile Station–Base Station Compatibility Standard for Dual-Mode Wideband Spectral System)[1]. This standard supports a direct sequence spread spectrum technology which is a form of CDMA technology with a shared frequency bandwidth of 1.23 MHz. In theory, as many as 64 mobile subscribers can share the frequency spectrum, although interference considerations can limit the maximum number of subscribers to a number that is substantially lower than this maximum value.

The intent of TR45.5 is to enable a service provider to add capacity to an existing analog cellular system. Consequently, TIA IS-95A supports a dual-mode mobile station for both CDMA and analog operation. Dual-mode operation allows a service provider to configure areas of CDMA/analog operation abutting areas of analog-only operation. An IS-95 compatible mobile station searches for a CDMA pilot. If CDMA operation is not detected by the mobile station, the mobile station may search for an analog overhead channel.[2] Furthermore, the mobile station may switch from CDMA to analog operation during a call if a CDMA to analog handoff command is issued by the base station to the mobile station.[3]

The analog capabilities in TIA IS-95 are built upon the analog capabilities of EIA 553 (now TIA IS-91) and TIA IS-54 with a few exceptions. For example, TIA IS-95 does not support the concept of secondary control channels since these channels could interfere with the spectrum assignment of another 1.25-MHz CDMA carrier. Also, TIA IS-95 supports a new control message on the analog overhead channel, which indicates the availability of CDMA, thus facilitating the migration of the mobile station to CDMA operation.

With the availability of a new frequency spectrum (1.8–2.0 GHz) for personal communications systems, TR46 adopted Standards SP-3384

---

1. This standard was later replaced with version IS-95A in 1994. In addition, a Technical Support Bulletin (PN-3570) was issued in 1995 to support the 14.4-kbps physical layer, Service Configuration and Negotiation, and PCS interaction. The title of this TSB is "Telecommunications System Bulletin: Support for 14.4 kbps Data Rate and PCS Interaction for Wideband Spread Spectrum Cellular Systems."

2. TIA IS-95 is purposely nebulous on this issue. Rather, the exact action is dependent upon the manufacturer. This ambiguity allows a mobile manufacturer to implement mobiles with some product differentiation.

3. CDMA standards do not support an analog-to-CDMA handoff. Such a handoff is technically complex and would require a very long blank-and-burst message on the analog voice channel. This type of handoff would cause a significant degradation of the voice quality during such handoffs. This capability is being studied by TR45.5 but is not considered a high priority.

that expanded upon IS-95. TR46 was formed by the TIA to formulate standards that addressed the environment encountered in this new radio spectrum and supported services that would attract customers. TR46 studied a number of technologies, one of which was CDMA. A number of new concepts were introduced in SP-3384 (e.g., service negotiation and a 14.4-kbps traffic channel). Some of these concepts were, in turn, adopted by TR45.5 and incorporated in standards TIA IS-95A and TSB PN-3570. At this writing, SP-3384, IS-95A, and TSB PN-3570 specify the basis for CDMA operation in North America. The remainder of this chapter will center around the basic concepts of these standards.

3.3 LAYERING CONCEPTS

TIA IS-95A supports the functionality of the physical layer, the data link layer, and the network layer (see chapter 5 for more details on the various layers). However, TIA IS-95A is not separately structured for each of these layers; rather, the specifications for these layers are somewhat interlaced.

In TIA IS-95A (refer to sections 6.1.1.1–6.1.3.2 and 7.1.1.1–7.1.3.5 of TIA IS-95A) the responsibility of the physical layer (layer 1) is to transmit raw bits over a communications channel (i.e., the sending and receiving of 1s and 0s over the radio channel). This topic will be discussed in chapter 5.

The task of the data link layer (layer 2) is to ensure that the raw transmission is transformed to be error-free for the network layer (refer to sections 6.1.3.2–6.2.2.3 and 7.1.3.5–7.1.3.5.11.4 of TIA IS-95A). The data link layer partitions the transmission into data frames every 20 ms, transmits the frames sequentially, error encodes, error detects, acknowledges the reception of frames on either side, may request for retransmission if an error is detected, and retransmits if requested by the other side. If a voice frame is detected as being in error, retransmission is not practical since the next voice frame must be sent in real-time. Sections 6.6.4.1.3 and 7.6.2.1.4 of TIA IS-95A present the acknowledgment procedures on the forward and reverse directions, respectively. In each signaling message, layer 2-related fields are included in order to detect duplicate messages, respond to acknowledgments, and request acknowledgments. Three related fields are ACK_SEQ (3 bits), MSG_SEQ (3 bits), and ACK_SEQ (1 bit) (See table 3.1).

The MSG_SEQ indicates the sequence number of the transmitted signaling message. If either the mobile station or the base station

Table 3.1 Layer 2 Message Fields. (Reproduced under written permission of the copyright holder [TIA].)

Field	Length, bits
MSG_TYPE	8
ACK_SEQ	3
MSG_SEQ	3
ACK_REQ	1
Remainder of message	

requires acknowledgment of a transmitted signaling message, the ACK_SEQ field is set to 1. Generally, the acknowledgment is included in a signaling message on the forward and reverse traffic channels. If no such signaling message needs to be sent and an acknowledgment of a received message is required (i.e., ACK_SEQ = 1), a Mobile Station Acknowledgment Order or Base Station Acknowledgment Order is sent on the reverse or forward traffic channels, respectively. If four or more messages that require acknowledgment are not acknowledged, additional messages that require acknowledgment will not be sent until an acknowledgment is received.

The network layer (layer 3) accepts messages from the source host, converts the messages into packets, and directs the packets toward the destination. Consequently, the network layer does routing of packets and controls the congestion of packets. In EIA/TIA IS-95A, the network layer corresponds to call processing, which will be discussed in section 3.4.

Other CDMA standards build on the three lower layers defined in TIA IS-95A. For example, the standardized version of short message service (SMS)[4] builds on top of IS-95A to provide an SMS relay layer and an SMS transport layer. The SMS teleservice layer interfaces with the SMS transport layer (see chapter 10).

3.4 CALL PROCESSING

As mentioned in section 3.3, call processing is situated at the network layer. The main responsibility of call processing is to accept commands from the source host (which is an IS-96A vocoder if the value of the ser-

---

4. TIA IS-637, Short Message Services for Wideband Spread Spectrum Cellular Systems.

vice option is 1), convert the commands into messages, direct the messages toward the destination, reconvert the messages to commands, and present the commands to the destination (which is an IS-96A vocoder). Consequently, call processing must be able to set up a call from the CDMA mobile station to the destination, which may be a telephone connected through the Public Switched Telephone Network (PSTN), or to another mobile station (analog, CDMA, TDMA, or GSM). Additionally, the cellular system must be able to transfer packets correctly as the mobile station moves from base station to base station during a call.

The call model adopted by TIA IS-95A is essentially the same call model as in the current analog standard EIA-553 (Mobile Station–Land Station Compatibility Specification). This call model supports a call from one calling directory number to one called directory number. Multiple simultaneous calls are not defined, and support is outside the realm of IS-95A.

The network layer also performs the tasks of power control,[5] mobile station lockout,[6] radio channel control,[7] and mobile station control.[8]

As with EIA-553, a mobile station can be addressed by the mobile station identification number (MIN), which is a transformation of the mobile station's directory number. However, with international operation, the MIN may not be unique, and thus the international mobile station identification (IMSI) has been introduced into TIA IS-95A. IMSI consists of 15 digits, while the MIN has 10 digits. The potential problem with identifying the mobile station with the MIN is that 10 digits are not sufficient to guarantee unique addressing for international roaming. Such a situation may occur in the near future if a homed-Mexican mobile

5. Power control is the function of the base station. The transmitted RF power of both the mobile and the base station are adjusted so that the frame error rate on the forward traffic channel and the reverse traffic channel is within the desired error rate (e.g., 1 percent).

6. If the mobile station does not properly respond to the power control commands from the base station, the base station may issue a command to the mobile to prevent further CDMA access by the mobile station. This capability is a safety mechanism to prevent "rogue" mobile stations from overloading the CDMA system and thus severely limiting the traffic capacity.

7. Radio channel control is the responsibility of the base station in which a CDMA channel is assigned to a call for either setup or handoff.

8. The base station issues commands to the mobile station in order to affect the mobile station's call state. Examples are paging, alerting (ringing), and releasing the call.

station roams into the United States. Also, the mobile station may be addressed by the electronic serial number (ESN).[9]

Call processing of an IS-95A mobile station is partitioned into four states (see fig. 3.1), which are further partitioned into substates. The four states are

- Mobile station initialization state,
- Mobile station idle state,
- System access state, and
- Mobile station control on the traffic channel state.

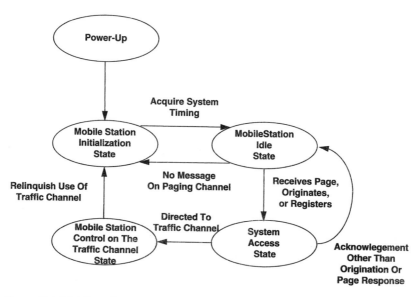

Figure 3.1 Mobile call processing states. (Reproduced under written permission of the copyright holder [TIA].)

3.4.1 Mobile Station Initialization State

The Mobile Station Initialization State consists of four substates as shown in Figure 3.1. In the *system determination substate*, the mobile station selects the cellular system. If the mobile station cannot find a suitable CDMA system, the mobile station may acquire an analog system. In such a case, the mobile station will follow a state model based

9. Messages on the paging channel that cannot use the ESN for addressing are Page Message and General Page Message.

upon the model that is specified in EIA-553. The mobile station acquires the pilot channel in the *pilot channel acquisition substate*. Next, the mobile station executes the *sync channel acquisition substate* in which the mobile station obtains the system configuration and timing information from the sync channel (see table 3.2).

Table 3.2 Sync Channel Message. (Reproduced under written permission of the copyright holder [TIA].)

Field	Length, bits
MSG_TYPE (00000001)	8
P_REV	8
MIN_P_REV	8
SID	15
NID	16
PILOT_PN	9
LC_STATE	42
SYS_TIME	36
LP_SEC	8
LTM_OFF	6
DAYLT	1
PRAT	2
RESERVED	3

This information is contained in the sync channel message, which provides such information as the system identification (SID) and network identification (NID) of the acquired system, system time, and the PN offset of base station. The final substate is the *timing change substate*. The mobile station adjusts its timing in preparation for receiving the paging channel, determines the primary paging channel, and initializes its registration process (but does not attempt registration).

3.4.2 Mobile Station Idle State

After completing the timing change substate, the mobile station enters the mobile station idle state. In this state, the mobile station monitors the paging channel, performs registration if required, and executes idle handoffs.

TIA IS-95A uses the term *idle handoff* to indicate that the mobile station has started to monitor a paging channel associated with a different base station than was previously monitored. Note that a mobile station monitors the paging channel of one base station; in other words, a soft handoff is not applicable in this state. The mobile station follows the registration procedures as will be discussed in section 3.5.

A number of tasks are performed by the mobile station while monitoring the paging channel. These tasks are

- Responding to overhead information,
- Responding to pages (mobile termination),
- Responding to mobile station orders,
- Initiating mobile originations,
- Powering down the mobile station, and
- Providing optional support of message transmission.

Both mobile termination and mobile origination are discussed in chapter 7 and will not be covered in this chapter. However, in this state, the mobile station determines if an origination, page response, or a registration is to be generated. The mobile station will access the system with the appropriate access channel message only when it is in the system access state.

Six overhead messages can be transmitted on the paging channel by the base station:

- Systems Parameters message,
- Neighbor List message,
- CDMA Channel List message,
- Extended System Parameters message,
- Access Parameters message, and
- Global Service Redirection message.

The System Parameters message will be discussed in section 3.7. This message also provides information for monitoring pilots and for controlling power. The Neighbor List message provides information about updating the mobile station's neighbor list. This list is used to assist with idle handoffs. The CDMA Channel List message indicates the paging channels that are supported by the base station being monitored by the mobile station. The Access Parameters message defines the parameters used by the mobile station when transmitting to the base station on the access channel (e.g., attempting registration or mobile origination). Finally, the Global Redirection message instructs the mobile station to move to another CDMA band or move to analog service.

3.4.3 System Access State

The mobile station will transition from the mobile station idle state to the system access state if the mobile station is required to send a message on the access channel. This corresponds to the mobile station sending the following messages:

- Origination message (mobile station origination attempt state),
- Page Response message (page response substate),
- Response to an order (mobile station order/message response substate),
- Data Burst message (mobile station message transmission substate), and
- Registration message (registration access substate).

The access procedures are discussed in detail in chapter 7. However, before the mobile station attempts to access the base station, the mobile station will monitor the paging channel until the access parameters have been updated (update overhead information substate), (see fig. 3.2).

3.4.4 Mobile Station Control on the Traffic Channel State

The mobile station is directed to the mobile station control on the traffic channel state if it receives a Channel Assignment message and the mobile station is currently in the access state. This state consists of the following five substates (see fig. 3.3):

- Traffic channel initialization substate,
- Waiting for order substate,
- Waiting for mobile station answer substate,
- Conversation substate, and
- Release substate.

In the *traffic channel initialization substate*, the mobile station successfully receives the base station's transmission on the forward traffic channel and begins transmitting on the reverse traffic channel. If the associated call is mobile-terminated, the mobile station goes into the

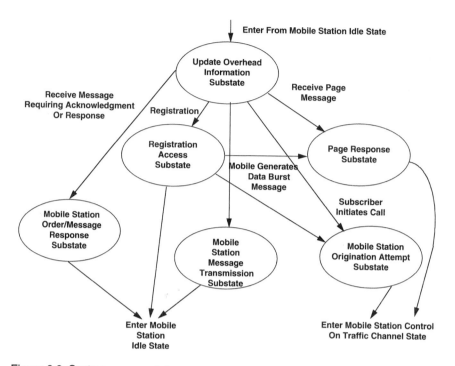

Figure 3.2 System access state. (Reproduced under written permission of the copyright holder [TIA].)

waiting for order substate and waits for an alert with information message (i.e., the mobile station is instructed to ring in order to alert the mobile subscriber). After ringing has been initiated for a mobile-terminated call, the mobile station waits for the mobile subscriber to "go off-hook," and the mobile station consequently moves into the *waiting for mobile station answer substate*. The mobile station transmits and receives traffic packets in the *conversation substate*. Either the mobile station or the base station can initiate a call release, corresponding to the *release substate*.

This section only outlines the mobile station's operation in the mobile station control on the traffic channel state. Chapters 6 and 7 will provide greater detail in the discussion of CDMA messaging and call flows.

3.4.5 Base Station Call States

TIA IS-95A does not invoke as detailed specifications for the base station as for the mobile station. This philosophy is consistent with EIA-553. Needless to say, the base station must operate with the mobile sta-

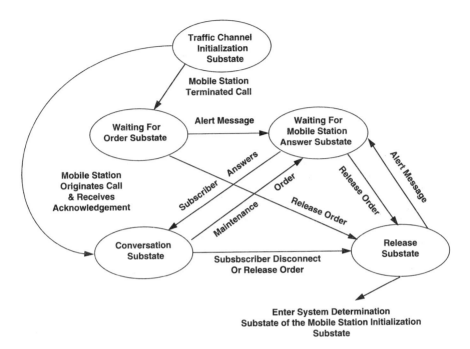

Figure 3.3 Mobile station control on the traffic channel state. (Reproduced under written permission of the copyright holder [TIA].)

tion; however, the manufacturers of base stations are offered more opportunity to differentiate among themselves.

TIA IS-95A defines only the traffic channel processing state, which corresponds to the mobile station control on the traffic channel state.

3.5 SERVICE CONFIGURATION AND NEGOTIATION

Service configuration is the common set of attributes used by the base station and the mobile station for interpreting and building frames on the forward and reverse traffic channels. *Service negotiation* is the process by which a mutually acceptable service configuration is agreed upon by both the base station (network) and the mobile station.

With the variety of services that can be supported on the CDMA air interface, TSB PN-3570 and SP-3384 allow the network and the mobile station to communicate with the appropriate configuration on the forward and reverse traffic channels. The configuration is determined by

- Service option,
- Service option connection reference,

- Forward traffic type, and
- Reverse traffic type.

The *service option* identifies the type of service being supported on a given connection. For example, different services correspond to different vocoder algorithms. TIA TSB-58 documents the 16-bit assignment for each standardized service type (see table 3.3).

Table 3.3 Service Options for CDMA. (Reproduced under written permission of the copyright holder [TIA].)

Service Option Number	Type of Service
1	Basic variable rate voice service (8 kbps)
2	Mobile station loopback (8 kbps)
3	Enhanced variable rate voice service (8 kbps)
4	Asynchronous data service
5	Group 3 facsimile
6	Short message service
7	Packet data service (Internet)
8	Packet data service (CDPD)
9	Mobile station loopback (13 kbps)
10	STU-III service option
11–32,767	Reserved for standard service options
32,768–65,535	Reserved for proprietary service options

Only the lower 32,768 values can be assigned to standardized services. In addition, the higher 32,768 values can be assigned to manufacturers for proprietary services, thus allowing some service differentiation among manufacturers. The *forward traffic type* specifies the physical link from the base station to the mobile station, while the *reverse traffic type* defines the physical link from the mobile station to the base station. Service configuration permits different configurations in the reverse and forward directions. While configuration asymmetry is not useful for voice services, it is very amenable to data services for which data rates are different in the forward and reverse directions. In order to specify the physical link, both the data rate and the multiplex option are needed. CDMA standards currently support two data rates: 9.6 and 14.4 kbps. These values correspond to the raw data rates and include both the transmitted information and error protection. Thus, 8.6 and 13.35 kbps of information can be transmitted on the respective data rates.

The multiplex option enables two logical channels to be transmitted on the same physical traffic channels. The primary channel is usually configured, while the second logical channel, called the *secondary traffic channel*, is logically superimposed if the multiplex option is equal to 2. If the multiplex option is equal to 1, only the primary traffic channel is configured. It should be noted that the multiplex options can be different for the forward and reverse directions.

Multiple service connections may be established on the same physical traffic channel (as specified by the rate set and multiplex option). Even though the call model as adopted by TIA IS-95A does not support multiple call appearances, multiple connections may be simultaneously established under certain configurations. For example, two simultaneous connections can be supported on the same physical link if connection 1 uses the primary traffic channels in the forward and reverse direction and connection 2 uses the secondary traffic channels in the forward and reverse directions. As many as four connections can be simultaneously established if each connection is unidirectional and uses one traffic channel type (i.e., either the primary or secondary channel). If, however, each channel is terminated at different end points, then a multiple dialing arrangement is required since the CDMA call model does not cover such scenarios.

3.6 CONCEPT OF SYSTEM IDENTIFICATION AND NETWORK IDENTIFICATION

A *network*, as identified by the network identification (NID), can selectively provide CDMA service within a CDMA system, as identified by the system identification (SID).

A network is thus a subset of a CDMA system. If a base station indicates that its associated NID = 0 in the System Parameters message, which is broadcast on the paging channel as an overhead message, then that base station should not be considered as part of the specific network.

The mobile station contains a list of SID,NID pairs in the number assignment module (NAM). The mobile station must have at least one SID,NID pair. If the stored SID,NID pair has the NID value equal to 65,535 ($2^{16}-1$), then the mobile station will consider the associated SID, regardless of the NID, as the home system.

As with EIA-553, a cellular system is identified by the systems identification, which is represented by a 15-bit field. However, TIA IS-95A introduces an additional dimension for defining a network with the concept of the network identification, which is represented by a 16-bit field. Thus, a network is uniquely specified by the pair SID,NID (see fig. 3.4).

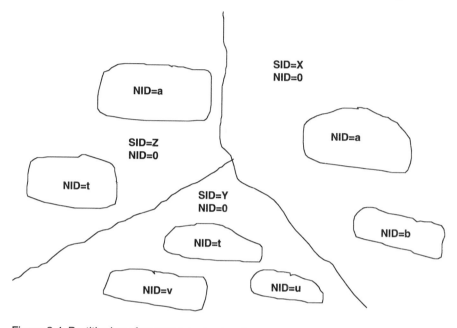

Figure 3.4 Partitioning of systems and networks. (Reproduced under written permission of the copyright holder [TIA].)

The mobile station's NAM contains at least one SID,NID pair and, thus, may have multiple pairs to designate the mobile station's home network. If the SID,NID pair transmitted in the System Parameters message is not included in the mobile station's NAM, the mobile station will consider itself as a roamer in the given network (i.e., it is located in a foreign NID). Similarly, if the SID is not contained in one of the NAM's pairs, the mobile station determines that it is a roamer in the given system (i.e., the mobile station is in a foreign SID). If the NID equals all zeros in the System Parameters message, then the broadcasting base station is not included in a specific network. In other words, all mobiles may access that base station.

3.7 REGISTRATION

Registration is a process by which a mobile station informs the network about pertinent status parameters. Practical registration parameters are relevant to location, time, and activation status. In the following discussion on registration, the base station actions are presented. When a mobile station registers, the mobile station will send a registration message to the base station (see table 3.4).[10]

Table 3.4 Content of Mobile Station Registration Message. (Reproduced under written permission of the copyright holder [TIA].)

Field	Length, bits
MSG_TYPE (00000001)	8
ACK_SEQ	3
MSG_SEQ	3
ACK_REQ	1
VALID_ACK	1
ACK_TYPE	3
MSID_TYPE	3
MSID_LEN	4
MSID	8 x MSID_LEN
AUTH_MODE	2
AUTHR	0 or 18
RANDC	0 or 8
COUNT	0 or 6
REG_TYPE	4
SLOT_CYCLE_INDEX	3
MOB_P_REV	8
SCM	8
MOB_TERM	1
RESERVED	6

10. A number of fields are not directly relevant to the registration process. The AUTH_MODE, AUTHR, RANDC, and the COUNT fields are applicable to authentication. The MSID_TYPE, MSID_LEN, and the MSID fields identify the addressing type of the mobile (e.g., the IMSI or the ESN of the mobile). The SLOT_CYCLE_INDEX and SLOTTED_MODE fields are used so the mobile can be paged at certain times, thus reducing the battery consumption. The MOB_P_REV and EXT_SCM fields identify the mobile station's hardware and software.

TIA IS-95A supports autonomous registration, ordered registration, and parameter-change registration. With *autonomous registration,* a mobile station will register without an explicit command from the base station. Autonomous registration will be discussed in detail later in this section. TIA IS-95A uses the term *implicit registration,* which is nothing more than the base station receiving an origination message or a page response message; hence, the mobile station's location status is known to the cellular system. A mobile station attempts a *parameter-change registration* if specific mobile station parameters are modified and if PARAMETER_REG equals 1 in the System Parameters message (see table 3.5).[11]

The relevant mobile station parameters that result in a parameter change registration are the preferred slot cycle index (SLOT_CYCLE_-INDEX), station class mark (SCM), and call termination enable indicators. Additionally, a parameter-change registration will be attempted if the SID,NID pair that is contained in the System Parameters message does not match any pair stored in the mobile station's SID_NID_LISTs. The topic of SID and NID and the relationship to roaming was covered earlier. If the PARAMETER_REG field = 1 in the System Parameters message, parameter-change registration is attempted even if autonomous registration is not enabled.[12]

The mobile station responds with an *ordered registration* if ordered by the base station with a Registration Request Order on the paging channel. This procedure should not be used normally since the capacity

---

11. T_ADD, T_DROP, T_COMP, and T_TDROP set the thresholds for possible handoffs and are discussed in chapter 9. SRCH_WIN_N, SRCH_WIN_R, and SRCH_WIN_A are parameters used by the mobile station when it searches for candidate pilots. The mobile station sends a Power Measurement Report message based upon conditions set by PWR_REP_THRES, PWR_REP_FRAMES, PWR_THRES_ENABLE, PWR_PERIOD_ENABLE, and PWR_REP_DELAY. This message supports the control of the mobile station's transmitted power.

12. One necessary condition for autonomous registration to be active is REG_ENABLEDs = YES. This will occur if one of the following logical statements is true:

The mobile is not roaming, HOME_REG field = 1 in the System Parameters message, and MOB_TERM_HOME = 1 in the mobile station's NAM.

The mobile is a foreign NID roamer, FOR_NID_REG = 1 in the System Parameters message, and MOB_TERM_FOR_NID = 1 in the mobile station's NAM.

The mobile is a foreign SID roamer, FOR_SID_REG = 1 in the System Parameters message, and MOB_TERM_FOR_SID = 1 in the mobile station's NAM.

Table 3.5 Content of System Parameters Message Transmitted by a Base Station.
(Reproduced under written permission of the copyright holder [TIA].)

Field	Length, bits	Field	Length, bits
MSG_TYPE (00000001)	8	BASE_LAT	22
PILOT_PN	9	BASE_LONG	23
CONFIG_MSG_SEQ	6	REG_DIST	11
SID	15	SRCH_WIN_A	4
NID	16	SRCH_WIN_N	4
REG_ZONE	12	SRCH_WIN_R	4
TOTAL_ZONES	3	NGHBR_MAX_AGE	4
ZONE_TIMER	3	PWR_REP_THRESH	5
MULT_SIDS	1	PWR_PERIOD_FRAMES	4
MULT_NIDS	1	PWR_THRESH_ENABLE	1
BASE_ID	16	PWR_PERIOD_ENABLE	1
BASE_CLASS	4	PWR_REP_DELAY	5
PAGE_CHAN	3	RESCAN	1
MAX_SLOT_CYCLE_INDEX	3	T_ADD	6
HOME_REG	1	T_DROP	6
FOR_SID_REG	1	T_COMP	4
FOR_NID_REG	1	T_TDROP	4
POWER_UP_REG	1	EXT_SYS_PARAMETER	1
POWER_DOWN_REG	1	RESERVED	2
PARAMETER_REG	1	GLOBAL_REDIRECT	1
REG_PRD	7		

of the paging channel is reduced and since autonomous registration procedures should be adequate.[13]

The mobile station will automatically attempt *autonomous registration* if the mobile station determines that a designated event has

---

13. IS-95A also refers to traffic channel registration as a form of registration. However, it is really a notification by the base station that the mobile is registered when the mobile is on the traffic channel (i.e., a CDMA call has been established). In such cases, the base station will send a Mobile Station Registered message.

occurred and that the particular type of autonomous registration is enabled by the cellular system. The cellular system enables (allowed forms of) autonomous registration by configuring the appropriate fields in the system parameters message.

TIA IS-95A supports the following types of autonomous registration:

- Power-up,
- Power-down,
- Timer-based,
- Zone-based, and
- Distance-based.

The TIA IS-95A standard allows *power-up registration* and *power-down registration* to be used independently of each other. However, the service provider most likely will use power-down registration only if power-up registration is activated. Power-up registration is performed when the battery power to the mobile station is activated. To prevent multiple registrations when the power is quickly turned on and off, the mobile station delays any action for 20 seconds before registering after power turn-on. The base station activates power-up registration by setting the POWER_UP_REG field in the System Parameters message to 1. Power-down registration is executed by the mobile station when the mobile subscriber deactivates the mobile station by turning off the mobile station. In such cases, if the POWER_DOWN_REG field is set to 1, and if the mobile station was previously registered in the given service area, the mobile station will not power down until completing a registration attempt.

Timer-based registration is a variation of autonomous registration that is supported in EIA-553 standards. A comparison is outside the scope of this writing. However, TIA IS-95A resolves a number of related deficiencies. With timer-based registration, the mobile station registers at regular intervals. The mobile station maintains the timer. Timer-based registration is activated by the base station if REG_PRD is not equal to 0000000. If so, the maximum count of the timer is $2^{\text{REG\_PRD}/4}$, where REG_PRD is the registration period. This timer is incremented once every 80 ms and is derived from the paging channel slot counter (note that each time slot is 80 ms in duration). When the timer reaches the maximum count and timer-based registration is enabled, the mobile station will generate a registration message, and the timer is reset. This timer is also reset on power-up, after each successful autonomous registration and after implicit registration. When the mobile station powers up, the timer is set to a random number, thus randomizing a mobile station with respect to other mobiles.

Zone-based registration is the fourth type of autonomous registration. This presupposes that the service provider groups base stations into zones. Each base station broadcasts its zone number. Different zone numbers may be associated for sectors of a given base station. The zone identification associated with the base station is contained in the 12-bit field REG_ZONE of the System Parameters message. If zone-based registration is disabled, the TOTAL_ZONES field of this message is set to 000. If the mobile station determines that the zone number has changed (by monitoring the System Parameters message on the paging channel), the new zone number is not in the stored zone list, and zone-based registration is enabled, the mobile station will register by sending a registration message on the access channel.

The mobile station will store a list of zones in which the mobile station has registered. Each registered zone corresponds to an entry in the ZONE_LIST, which consists of the zone number (REG_ZONE) and the SID,NID pair associated with the zone. The maximum number of entries is determined by the TOTAL_ZONES, which is controlled by the base station in the System Parameters message. If the associated timer has expired (the duration is determined by the ZONE_TIMER in the System Parameters message), the entry will be deleted.

Distance-based registration is the last type of autonomous registration. For distance-based registration to be active, the REG_DIST field in the System Parameters message must be nonzero. If so, the mobile station will register each time the mobile station travels this distance after the previous registration. The distance is determined by

$$\frac{\sqrt{(\Delta lat)^2 + (\Delta long)^2}}{16} \qquad (3.1)$$

where Δlat is calculated from the BASE_LAT and $\Delta long$ is calculated from the BASE_LONG as contained in the System Parameters message.

The unit of BASE_LAT and BASE_LONG is 0.25 angular seconds, which corresponds to approximately 25 feet. Δlat is the difference in the current base station's latitude with respect to the latitude of the base station in which the mobile station last registered. $\Delta long$ is the corresponding difference in the longitude.

A system provider may use none or all types of autonomous registration. Most likely, a subset of available types of registration will be supported by a cellular system. For example, the service provider may
choose to support power-up, power-down, and zone-based registration
simultaneously. However, it is doubtful that both zone-based registration
and distance-based registration would be simultaneously activated.

3.8 WIDEBAND CDMA STANDARDS

The previous sections discussed current CDMA standards (i.e., TIA IS-
95A and TIA PN-3570) that support a frequency bandwidth of 1.25 MHz.
Even though the titles of the corresponding standards imply wideband
CDMA, only standard J-STD-007 (PCS-1900 Air Interface Wideband
PCS Standard) currently supports larger frequency bandwidths of 5, 10,
and 15 MHz. To facilitate discussion, the former set of standards is identified as CDMA, while the latter standard is referenced as W-CDMA.

In addition, there are differences in the network layer (layer 3).
Unlike CDMA, W-CDMA incorporates a call model based upon the ISDN
call model. A call reference is included in call establishment messages.
Thus, it is possible for multiple call appearances (multiple simultaneous
calls) to be implemented. W-CDMA does not support the complete suite
of registration types; only power-on, power-off, and zone-based registrations are defined. Like CDMA, W-CDMA supports the equivalent of soft
handoffs and hard handoffs. However, W-CDMA refers to these two types
of handoffs as Type A handoffs and Type B handoffs, respectively. For
more details on W-CDMA see chapters 5–7.

3.9 SUMMARY

The purpose of this chapter is to provide a survey of the CDMA standards specifying the air interface (i.e., the messaging between the base
station and the mobile station). Even though TIA IS-95A is not explicitly
structured as such, three protocol layers are implied: the physical, the
data link, and the network layers. The call model is basically the same
call model that is incorporated in the existing analog standards (EIA-
553) with four call states: mobile station initialization, mobile station
idle, mobile access, and mobile station control on the traffic channel.
Each of these states is further partitioned into substates. For each call,
either the mobile station or the network may negotiate for a service type,
which is specified by the service option. In practice, the majority of calls
correspond to a default value such as basic variable rate voice service.

This chapter also discusses the types of autonomous registration, which supports the mobility function. CDMA operation supports power-up, power-down, time-based, zone-based, and distance-based registration.

3.10 REFERENCES

1. TIA/EIA IS-95A, "Mobile Station-Base Station Compatibility Standard for Dual-Mode Wideband Spread Spectrum Cellular System," March 1995.
2. TIA PN-3570 (TSB-74), "Telecommunications Systems Bulletin: Support for 14.4 Kbps Data Rate and PCS Interaction for Wideband Spread Spectrum Cellular Systems," October 1995.
3. TIA/EIA IS-96A, "Speech Service Option Standard for Wideband Spread Spectrum Digital Cellular System," May 1995.
4. TIA SP-3384, "Personal Station–Base Station Compatibility Requirements for 1.8 to 2.0 GHz Code Division Multiple Access (CDMA) Personal Communications Systems," March 1995.
5. TIA/EIA PN-3139 (to be published as TSB-58), "Telecommunications System Bulletin: Administration of Parameter Value Assignments for TIA/EIA Wideband Spectrum Standards," August 1995.
6. Tanenbaum, A.S., "Computer Networks," Prentice Hall, Englewood Cliffs, NJ, 1981.
7. EIA-553, "Mobile Station–Land Station Compatibility Specification," September 1989.
8. PCS-1900 Air Interface Proposed Wideband CDMA PCS Standard, J-STD-007, T1 LB-461.

System Architecture
for Wireless
Communications

4.1 INTRODUCTION

A wireless system, whether for cellular operation (i.e., 850 MHz), or for personal communication systems operation (i.e., 1.8 GHz), must support communication with the mobile station and interact with the Public Switched Telephone Network. As the mobile station changes its location during a call, the wireless system must ensure that the connection between the mobile station and the PSTN is maintained.

A wireless system consists of discrete logical components that may be either discrete physical entities or physically located with another logical entity. It is necessary that these functional entities interact in order to coordinate operation. Such interaction is achieved by messaging over interfaces between two entities. If two functional entities are physically separate and if the interface is standardized, it is possible that the service provider can purchase products from different manufacturers. However, successful operation is not guaranteed since the associated standard often does not cover all facets of operation. This may necessitate testing between manufacturer's equipment to eliminate differences that jeopardize proper interaction.

This chapter discusses functional entities and the standardized interfaces between those entities that have been standardized by the wireless communication industry.

4.1.1 TR-45/TR-46 Reference Model

Key to the North American Systems is the use of a common reference model from the cellular standards group TR-45. When work started on PCS, the TR-46 standards group adopted the TR-45 reference model for PCS, but with some minor changes in the names of the elements. A second reference model has been proposed by T1P1, but it is similar to the TR-45/TR-46 model. The names of each of the network elements are similar and some of the functionality is partitioned differently between the models. The main difference between the two reference models is how mobility is managed. Mobility is the capability for users to place and receive calls in systems other than their home system. In the T1P1 reference model, the user data and the terminal data are separate; thus, users can communicate with the network via different mobile stations. In the TR-45/TR-46 reference model, only terminal mobility is supported. A user can place or receive calls at only one terminal (the one the network has identified as owned by the user). The T1P1 functionality is migrating toward independent terminal and user mobility, but all aspects of it are not currently supported. In chapter 11 we will discuss a wireless intelligent network (WIN) architecture and reference model that overcomes some of the mobility problems in the current architecture. Figure 4.1 shows the TR-45/TR-46 reference model.

4.1.2 Elements of the Reference Model

The main elements of the reference model follow:

- **Mobile Station (MS):** The MS terminates the radio path on the user side and enables the user to gain access to services from the network. The MS can be a stand-alone device or can have other devices (e.g., personal computers and fax machines) connected to it.
- **Base Station (BS):** The base station terminates the radio path and connects to the mobile switching center. The base station is often segmented into the BTS and the BSC:

    ✗ *Base Transceiver System (BTS):* The BTS consists of one or more transceivers placed at a single location and terminates the radio path on the network side. The BTS may be either co-located with a BSC or independently located.

    ✗ *Base Station Controller (BSC):* The BSC is the control and management system for one or more BTSs. The BSC

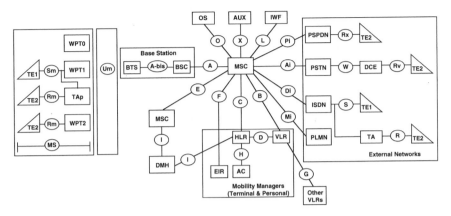

Figure 4.1 TR-45/TR-46 reference model. (Reproduced under written permission of the copyright holder [TIA]. At time of publication, the standard that contains this figure was not finalized. Check with TIA for the correct version.)

exchanges messages with both the BTS and the MSC. Some signaling messages may pass through the BSC transparently.

- **Mobile Switching Center (MSC):** The MSC is an automatic system that interfaces the user traffic from the wireless network to the wireline network or other wireless networks. The MSC functions as one or more of the following:

  ✗ *Anchor MSC*—the first MSC providing radio contact in a call.

  ✗ *Border MSC*—an MSC controlling BTSs adjacent to the location of a mobile station.

  ✗ *Candidate MSC*—an MSC that could possibly accept a call or a handoff.

  ✗ *Originating MSC*—the MSC directing an incoming call toward a mobile station.

  ✗ *Remote MSC*—the MSC at the other end of an intersystem handoff trunk.

  ✗ *Serving MSC*—the MSC currently providing service to a call.

  ✗ *Tandem MSC*—an MSC providing only trunk connections for a call in which a handoff has occurred.

  ✗ *Target MSC*—the MSC selected for a handoff.

  ✗ *Visited MSC*—an MSC providing service to the mobile station.

- **Home Location Register (HLR):** The HLR is the functional unit used for management of mobile subscribers by maintaining all subscriber information (e.g., electronic serial number, directory number, international mobile station identification, user profiles, and current location). The HLR may be co-located with a MSC as an integral part of the MSC or may be independent of the MSC. One HLR can serve multiple MSCs, or an HLR may be distributed over multiple locations.

- **Data Message Handler (DMH):** The DMH is used for collecting billing data and is described in chapter 11.
- **Visited Location Register (VLR):** The VLR is linked to one or more MSCs. The VLR is the functional unit that dynamically stores subscriber information (e.g., ESN, directory number, and user profile information) obtained from the user's HLR when the subscriber is located in the area covered by the VLR. When a roaming MS enters a new service area covered by an MSC, the MSC informs the associated VLR about the MS by querying the HLR after the MS goes through a registration procedure.
- **Authentication Center (AC):** The AC manages the authentication or encryption information associated with a individual subscriber. As of the writing of this book, the details of the operation of the AC have not been finalized. The AC may be located within an HLR or MSC or may be located independently of both.
- **Equipment Identity Register (EIR):** The EIR provides information about the mobile station for record purposes. As of the writing of this book, the details of the operation of the EIR have not been defined. The EIR may be located within a MSC or may be located independently of it.
- **Operations Systems (OSs):** The OSs are responsible for overall management of the wireless network. See chapter 10 for a full description of the OAM&P functions.
- **Interworking Function (IWF):** The IWF enables the MSC to communicate with other networks. See chapter 12 for details on interworking.
- **External Networks:** These communications networks include the Public Switched Telephone Network (PSTN), the Integrated Services Digital Network (ISDN), the Public Land Mobile Network (PLMN), and the Public Switched Packet Data Network (PSPDN).

The following interfaces are defined among the various elements of the system:

- **BS to MSC (A-Interface):** The interface between the base station and the MSC supports signaling and traffic (both voice and data).

A-Interface protocols have been defined using SS7, ISDN BRI/PRI, and frame relay.

- **BTS to BSC Interface (A_{bis}):** If the base station is segmented into a BTS and BSC, this internal interface is defined.

- **MSC to PSTN Interface (A_i):** This interface is defined as analog interface using either dual-tone multifrequency (DTMF) signaling or multifrequency (MF) signaling.

- **MSC to VLR (B-Interface):** This interface is defined in the TIA IS-41 protocol specification [4].

- **MSC to HLR (C-Interface):** This interface is defined in the TIA IS-41 protocol specification.

- **HLR to VLR (D-Interface):** This interface is the signaling interface between an HLR and a VLR and is based on SS7. It is currently defined in the TIA IS-41 protocol specification.

- **MSC to ISDN (D_i-Interface):** This is the digital interface to the PSTN and is a T1 interface (24 channels of 64 kbps) and uses Q.931 signaling.

- **MSC-MSC (E-Interface):** This interface is the traffic and signaling interface between wireless networks. It is currently defined in the TIA IS-41 protocol specification.

- **MSC to EIR (F-Interface):** Since the EIR is not yet defined, the protocol for this interface is not defined.

- **VLR to VLR (G-Interface):** When communications are needed between VLRs, this interface is used. It is defined by TIA IS-41.

- **HLR to Authentication Center (H-Interface):** The protocol for this interface is not defined.

- **DMH to MSC (I-Interface):** This interface is described in chapter 11.

- **MSC to the IWF (L-Interface):** This interface is defined by the inter-working function.

- **MSC to PLMN (M_i-Interface):** This interface is to another wireless network.

- **MSC to OS (O-Interface):** This is the interface to the operations systems. It is currently being defined in ATSI standard body T1M1 (see chapter 11).

- **MSC to PSPDN (P_i-Interface):** This interface is defined by the packet network that is connected to the MSC.

- **Terminal Adapter (TA) to Terminal Equipment (TE) (R-Interface):** These interfaces will be specific for each type of terminal that will be connected to a MS.
- **ISDN to TE (S-Interface):** This interface is outside the scope of PCS and is defined within the ISDN system.
- **Base Station to MS (U_m-Interface):** This is the air interface. Chapters 5–8 will discuss this interface in detail.
- **PSTN to DCE (W-Interface):** This interface is outside the scope of PCS and is defined within the PSTN system.
- **MSC to AUX (X-Interface):** This interface depends on the auxiliary equipment connected to the MSC.

4.2 STANDARDIZATION OF THE MSC-BS INTERFACE

North American Standards, until recently, have not addressed the standardization of the BS-MSC interface (the A-interface in the network reference model of fig. 4.1). However, the wireless service providers are experiencing explosive growth in North America and are consequently finding it necessary to purchase equipment from multiple equipment manufacturers. Thus, the wireless industry has pressed for standards specifying the A-interface. At this time, however, the BTS-BSC interface (i.e., the A_{bis} interface) is not being addressed. The TIA TR-45 Committee is currently developing standards for the A-interface. TIA IS-634 (MSC-BS interface for 800 MHz) has the following objectives:

- Develop the MSC-BS interface based on the TIA TR-45 network reference model;
- Partition the responsibility of functions provided between the base station and the mobile switching center without dictating specific implementation;
- Support North American air interface signaling protocols including EIA/TIA IS-95A;
- Support all the services offered to mobile subscribers operating under North American standards.

TIA IS-634 is partitioned into six major sections

- Functional Overview (IS-634.1),
- Call Processing and Supplementary Services (IS-634.2),
- Radio Resource Management (IS-634.3),
- Mobility Management, Authentication, and Privacy (IS-634.4),

- Layer 1 & 2 and Tererestrial Facility Management (IS-634.5), and
- Messages, Parameters, and Timer Definitions (IS-634.6).

TIA IS-634 defines MSC-BS messages, message sequencing, and mandatory timers at the base station and the mobile switching center. The discussion of this standard will be limited to the architectural impact of the MSC-BS interface rather than upon the associated message flow. In TIA IS-634, the base station is really the base station controller; thus, multiple BTSs may be connected to the base station. Also, TIA IS-634 refers to the BTS as the cell. Of course, the BTS and the BSC can be co-located.

Call processing, radio resource management, mobility management, and transmission facilities management are separate functions that are supported by the applications layer (see fig. 4.2).

The underlying transport mechanism for the applications layer is ISDN with the physical layer specified by ANSI T1.101, the message transfer part (MTP) specified by ANSI T1.111, and the signaling connection control part (SCCP) specified by ANSI T1.112. The physical interface supports one or more 1.544-Mbps digital transmission facilities, each providing twenty-four 56-kbps or 64-kbps channels. Each channel can be used for traffic or for signaling. The MTP and the SCCP support only signaling messages, whereas the physical layer supports both signaling messages and traffic messages. Traffic messages carry voice transmission. TIA IS-634 allows the transcoder (vocoder) to reside either at the

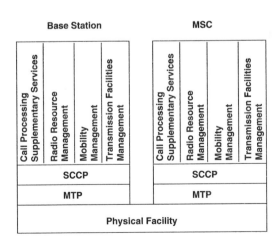

Figure 4.2 TIA IS-634 functions. (Reproduced under written permission of the copyright holder [TIA].)

base station or "very near" to the mobile switching center. In the first case, an entire DS0[1] (64 kbps) connection is required for each call, whereas the second case does not necessitate an entire DS0 connection.

At the applications layer, the call processing and mobility management function are connected between the mobile station and the mobile switching center, while the radio resource management and the transmission facilities management functions are connected between the base station and the mobile switching center. Accordingly, the base station application part (BSAP), which is the applications layer signaling protocol, is divided into two subapplications parts. The first is called the base station management application part (BSMAP). The BSMAP messages are sent between the base station and the mobile switching center. The second is the direct transfer application part (DTAP) in which messages are sent between the mobile station and the mobile switching center. The base station acts as a transparent conduit for DTAP messages. The base station merely maps the messages going to/from the mobile switching center into the appropriate air interface signaling protocol (e.g., TIA/EIA IS-95A). This approach simplifies the role of the base station for call processing and mobility management.

The base station associates the DTAP messages with a particular mobile station and call using a transaction identification. The BSAP messages are transferred over an SCCP connection. The DTAP and BSMAP layer 3 messages between the base station and the mobile switching center are contained in the user data field of SCCP frames. The data field is supported in connection request (CR), connection confirm (CC), released (RLSD), and data (DT) SCCP frames for mobile stations having one or more active transactions. The layer 3 user data field is partitioned into three components (see fig. 4.3):

- BSAP message header;
- Distribution data unit, which includes the Length Indicator and the Data Link Connection Identifier (DLCI)—only DTAP message; and
- Layer 3 Message.

The BSAP message header consists of the message discrimination and the data link connection identifier, which is applicable only for DTAP messages. The D-bit (bit 0) of the message discrimination octet is set to 1

---

1. A DS0 is a 64-kbps pulse code modulation (PCM) transport facility and is a single 64-kbps time slot on a T1 carrier.

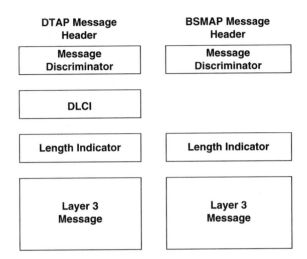

Figure 4.3 Layer 3 data field.

for a DTAP message and set to 0 for a BSMAP message. The distribution data unit consists of the length indicator octet, which gives the number of octets following the length indicator. The layer 3 message will be discussed in greater detail.

The DTAP messages apply only to mobility management and to call processing (including supplementary services) functions, whereas BSMAP messages are associated with radio resource management and call processing (to a lesser degree). Each DTAP message contains the protocol discriminator octet, which identifies the associated procedure (i.e., call control, mobility management, radio resource management, and facilities management). All DTAP and BSMAP messages are identified by the message type octet.

The remaining part of this section provides greater detail for each function supported by the BSAP.

4.2.1 Supported Architectural Configurations

TIA IS-634 makes a number of assumptions regarding the underlying CDMA architecture. The basic architecture is shown in figure 4.4.

The main entities are the mobile switching center (MSC), the transcoder (XC), the base station (BS), the base transceiver system (BTS), and the mobile station (MS). The MS is not shown in figure 4.4, but the MS communicates with the BTS over the air interface. The func-

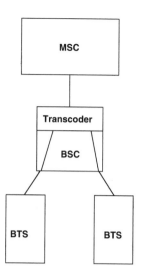

Figure 4.4 Basic CDMA architecture.

tions associated with the MSC and the MS were discussed in section 4.1. TIA IS-634 assumes that the base station is really the BSC. One or more BTSs are connected to the BSC.

The transcoder supports both voice coding (vocoder) and diversity reception. Diversity reception allows the transcoder to pick the "best" frame when multiple connections are established during a soft handoff. Diversity reception distinguishes CDMA technology from current digital technology. To be more specific, the XC is responsible for the following:

- Distribution of speech/data on the forward traffic channel to all BTSs associated with a call. During a soft handoff, multiple BTSs are simultaneously assigned to the call. The XC selects the best speech/data frame from all BTSs associated with the call on the reverse traffic channel. This implies that signal quality characteristics of the speech/data frame are provided to the transcoder.
- Decode QCELP[2] format to PCM format for voice frames sent on the reverse traffic channel. If the call is a data call, this task is bypassed.

2. QUALCOMM code-excited linear prediction (QCELP) is the CDMA speech processing algorithm that is specified in TIA IS-96A. The algorithm is based on code-excited linear prediction. See chapter 8 for more information.

- Decode PCM format to QCELP format for voice frames sent on the forward traffic channel. If the call is a data call, this task is bypassed.
- Rate adapt voice frames to fully use the transmission bandwidth of the assigned terrestrial circuits. This task is bypassed for data calls.
- Rate adapt compressed voice PCM format into a circuit switched subrate channel on a DS0 facility. One common compression approach is adaptative differential pulse code modulation (ADPCM). Compression uses the fact that voice activity is less than 100 percent of the total duration. Typically, the actual voice activity is approximately 50 percent.
- Provide a control capability of inserting blank and burst or dim and burst signaling into the voice transmission on the forward traffic channel.

The transcoder is considered as a logical part of the BS, although the transcoder can be physically located at the BS, or at the MSC, or somewhere between the BS and the MSC. The terrestrial facility connects the transcoder to the MSC. The terrestrial facility may be full-rate (56 kbps or 64 kbps), subrate, PSTN, bypass, or PSTN/bypass. If only the BS is associated with a call, the terrestrial facility is connected to the PSTN. If a call is configured for a soft handoff between two base stations, another terrestrial facility is required to connect the transcoder at the target BS with the transcoder at the source BS. Such a circuit is called a bypass facility. A bypass/PSTN facility is initially a bypass connection but subsequently is a PSTN connection. The connection between the transcoder and the BS is not addressed in TIA IS-634.

The transcoder and the BS may be physically co-located or externally connected by a full-rate or subrate facility if the transcoder is located near or at the MSC. The BSC may support multiple BTSs. Thus, if a call is in a soft handoff using only BTSs connected to a given BSC, no messaging between the MSC and the BS are necessary.

BTSs are uniquely identified by the cell global identification (CGI). The CGI is composed of four components:

- Mobile Country Code (MCC),
- Mobile Network Code (MNC),
- Location Area Code (LAC), and
- Cell Identity (CI).

TIA IS-634 supports addressing modes to identify a BTS by the CGI, or by the CI, or by a combination of the LAC, MCC, and MNC, or by the associated BS.

4.2.2 Call Processing and Supplementary Services

TIA IS-634 supports call setup (mobile origination and mobile termination) as well as supplementary services (e.g., call waiting) and call release. However, the support of handoffs during a call is considered as part of the radio resource management function, which will be discussed in section 4.3.2.

Most of the messages associated with call processing and supplementary services are DTAP messages. For these messages, the role of the base station is minimized since the base station "passes" the messages to the mobile station.

The initial BS-MSC message in the call setup procedure includes the mobile identity. The mobile identity can be the mobile identification number (MIN), the mobile station electronic serial number (ESN), or the international mobile subscriber identifier (IMSI). The identity type is selected by either the mobile station or the wireless network. For a mobile origination, the initial BS-MSC message is the connection management service request, and for a mobile termination, the initial BS-MSC message is the paging request. However, the initial message from the base station to the mobile switching center at call setup is an encapsulated DTAP message within a BSMAP message. The mobile switching center sends an assignment request message, which contains the terrestrial channel. Also, the mobile switching center may select the radio channel or provide channel parameters and allow the base station to choose the radio channel at the appropriate BTS.

Unique call processing for mobile-to-mobile calls are currently under study. Tables 4.1 and 4.2 list the messages defined by TIA IS-634 for call processing and supplementary services, respectively.

4.2.3 Radio Resource Management

Once a call has been established (i.e., call setup has been successfully completed), the base station is responsible for maintaining a reliable radio link between the mobile station and the base station. This responsibility requires that the base station perform the following tasks:

• Radio channel supervision,

Table 4.1 Call Processing Messages. (Reproduced under written permission of the copyright holder [TIA].)

Message Name	Direction	Message Type
CM Service Request	BS → MSC	DTAP
Paging Request	MSC → BS	BSMAP
Paging Response	BS → MSC	DTAP
Setup	BS ↔ MSC	DTAP
Emergency Setup	BS → MSC	DTAP
Alerting	BS ↔ MSC	DTAP
Call Confirmed	BS → MSC	DTAP
Call Proceeding	MSC → BS	DTAP
Connect	MSC → BS	DTAP
Connect Acknowledge	BS ↔ MSC	DTAP
Progress	MSC → BS	DTAP
Release	MSC ↔ BS	DTAP
Release Complete	MSC ↔ BS	DTAP
Assignment Request	MSC ↔ BS	BSMAP
Assignment Complete	BS → MSC	BSMAP
Assignment Failure	BS → MSC	BSMAP
Privacy Mode Command	MSC → BS	BSMAP
Privacy Mode Complete	BS → MSC	BSMAP
Clear Request	BS → MSC	BSMAP
Clear Command	MSC → BS	BSMAP
Clear Complete	BS → MSC	BSMAP

- Radio channel management, and
- Initiation and execution of handoffs.

The objective for each of these tasks is common for all radio technologies, although the actual implementation is dependent on the associated technology.

The support of soft handoffs is one capability that distinguishes CDMA from other access technologies. Thus, TIA IS-634 supports the procedures associated with soft handoffs. These procedures are

Table 4.2 Supplementary Service Messages. (Reproduced under written permission of the copyright holder [TIA].)

Message Name	Direction	Message Type
Send Burst DTMF[a]	BS ↔ MSC	DTAP
Send Burst Acknowledge	BS ↔ MSC	DTAP
Start DTMF	BS → MSC	DTAP
Start DTMF Acknowledge	MSC → BS	DTAP
Stop DTMF	BS → MSC	DTAP
Stop DTMP Acknowledge	MSC → BS	DTAP
Flash with Information	BS ↔ MSC	DTAP

a. Dual-Tone Multifrequency

- IS-95 Add Target Procedure,
- IS-95 Drop Target Procedure, and
- IS-95 Drop Source Procedure.

The source BS is the BSC, which controls the transcoder. If either a hard or soft handoff is to be configured with a target BTS that is connected to another BSC (target BS), then a handoff-required message is sent to the mobile switching center. The mobile switching center then sends a handoff request to the target BS. At the same time, only one target BS can be addressed.

Table 4.3 summarizes messages associated with radio resource management.

4.2.4 Mobility Management

Mobility management is implemented using DTAP messages. The purpose of the mobility management function is to support registration and deregistration of a mobile. In addition, this function encompasses authentication and voice privacy. Authentication includes authentication challenge and shared secret data (SSD) update. There is little differential impact upon the BS-MSC architecture in order to support this function.

Messages associated with mobility management are listed in table 4.4.

4.2.5 Transmission Facilities Management

The transmission facilities management function is responsible for the management of terrestrial circuits. *Terrestrial circuits* are transmission facilities that carry traffic (voice or data) and signaling information

Table 4.3 Handoff Messages. (Reproduced under written permission of the copyright holder [TIA].)

Message Name	Direction	Message Type
Strength Measurement Request	BS ↔ MSC	BSMAP
Strength Measurement Response	BS ↔ MSC	BSMAP
Strength Measurement Report	BS ↔ MSC	BSMAP
Handoff Required	BS → MSC	BSMAP
Handoff Request	MSC → BS	BSMAP
Handoff Request Acknowledge	BS → MSC	BSMAP
Handoff Failure	BS → MSC	BSMAP
Handoff Command	MSC → BS	BSMAP
Handoff Required Reject	MSC → BS	BSMAP
Handoff Commenced	BS → MSC	BSMAP
Handoff Complete	BS → MSC	BSMAP
Handoff Performed	BS → MSC	BSMAP
Soft Handoff Drop Target	BS ↔ BS	BSMAP
Soft Handoff Drop Source	BS ↔ BS	BSMAP

Table 4.4 Mobility Management Messages. (Reproduced under written permission of the copyright holder [TIA].)

Message Name	Direction	Message Type
Authentication Request	MSC → BS	DTAP
Authentication Reject	MSC → BS	DTAP
SSD Update Request	MSC → BS	DTAP
Base Station Challenge	BS → MSC	DTAP
Base Station Challenge Response	MSC → BS	DTAP
SSD Update Response	BS → MSC	DTAP
Location Updating Request	BS → MSC	DTAP
Location Updating Accept	MSC → BS	DTAP
Location Updating Reject	MSC → BS	DTAP
Parameter Update Request	MSC → BS	DTAP
Parameter Update Confirm	BS → MSC	DTAP

between the MSC and the BS. Furthermore, different facilities may carry traffic information from facilities carrying signaling information. Currently, TIA IS-634 does not address the facilities between the BS and the transcoder and between the BSC and BTSs. Each facility may be blocked/unblocked and allocated/deallocated by the transmission facilities management function. For digital technologies (e.g., CDMA), this function can disable the transcoders at both the originating end and the terminating end for mobile-to-mobile calls. This action eliminates the need for vocoder tandeming, which degrades the voice quality of a call. However, TIA IS-634 does not explicitly address this capability for calls spanning multiple BSs.

Table 4.5 summarizes the message types associated with transmission facilities management.

Table 4.5 Transmission Management Messages. (Reproduced under written permission of the copyright holder [TIA].)

Message Name	Direction	Message Type
Overload	MSC ↔ BS	BSMAP
Block	BS → MSC	BSMAP
Block Acknowledge	MSC → BS	BSMAP
Unblock	BS → MSC	BSMAP
Unblock Acknowledge	MSC → BS	BSMAP
Reset	BS ↔ MSC	BSMAP
Reset Acknowledge	BS ↔ MSC	BSMAP
Reset Circuit	BS ↔ MSC	BSMAP
Reset Circuit Acknowledge	BS ↔ MSC	BSMAP
Transcoder Control Request	MSC ↔ BS	BSMAP
Transcoder Control Acknowledge	MSC ↔ BS	BSMAP

4.3 SERVICES

With the reference model described previously, there are enough capabilities to support a wide range of telecommunications services over cellular or *personal communications systems*. Many of these services are similar to those of the wireline network; some are specific to the untethered approach that wireless provides. The services defined here are based on a mobile application part (MAP) that is supported by the IS-41 intersystem communications protocol. We discuss the services from the point of view of a CDMA (or wideband CDMA) phone, but the services provided by other air interfaces are the same. The main difference is how handoffs are handled in a CDMA system. We will therefore examine handoffs in detail.

4.3.1 Basic Services

The standards body T1P1 is in the process of defining basic call functions and supplementary services for PCS. Similarly, services for cellular systems are defined in TIA SP-2977. However, the objective of defining services is to provide transparency for the mobile subscriber regardless of the underlying technology of the serving cellular or PCS system.

The T1P1 Stage Two Service description [7] defines 15 basic services (information flows) that can be grouped as follows:

- Registration and deregistration functions support the process where an MS informs a personal communications system of its desire to receive service and its approximate location. These include

  ✗ Automatic registration,
  ✗ Terminal authentication and privacy (using private key cryptography),
  ✗ Terminal authentication and privacy (using public key cryptography),
  ✗ User authentication and validation,
  ✗ Automatic personal registration,
  ✗ Automatic personal deregistration,
  ✗ Personal registration, and
  ✗ Personal deregistration.

- The registration and deregistration process requires that an MS identify itself to the PCS network and requires that the PCS network communicate with the home PCS network to obtain security and service profile information.
- Roaming is the process where an MS registers and receives service in a personal communications system other than its home system.
- Call establishment, call continuation, and termination procedures include

  ✗ Call origination,
  ✗ Call delivery (call termination),
  ✗ Call clearing,

✘ Emergency (E911) calls, and

✘ Handoff.

4.3.2 Supplementary Services

Supplementary services are defined in IS-104 "Personal Communications Service Descriptions for 1800 MHz" (PN-3168). The IS-41 C specification defines them as those services that can be made available to users as they roam. Obviously, these services would also be available to users in their home systems. Additional services may be available in a specific home PCS or cellular system, but users would not necessarily have them available in other systems since no common set of procedures and protocols have been defined to support other services.

Supplementary services follow:

- **Automatic Recall** allows a wireless subscriber calling a busy number to be notified when the called party is idle and have the PCS network recall the number.
- **Automatic Reverse Charging (ARC)** allows a wireless subscriber to be charged for calls to a special ARC number. This service is similar to wireline 800 service in North America.
- **Call Hold and Retrieve** allows a wireless subscriber to interrupt a call and return to the call.
- **Call Forwarding (CF)—Default** represents the ability to redirect a call to an MS in three situations: unconditional, busy, and no answer. The MS call forwarding features build upon the MS call terminating capability. Under all these features, calls may be forwarded by the network to another mobile station or to a DN associated with a wireline interface. There are no additional information flows for MS-CF beyond the information flow for MS call terminating.
- **Call Forwarding—Busy** permits a called PCS subscriber to have the system send incoming calls addressed to the called personal communications subscriber's personal number to another personal, terminal, or directory number when the PCS subscriber is engaged in a call. With personal call forwarding—busy activated, a call incoming to the PCS subscriber will be automatically forwarded to the forward-to number whenever the PCS subscriber is already engaged in a prior call.
- **Call Forwarding—No Answer** permits a called PCS subscriber to have the system send all incoming calls addressed to the called PCS

subscriber's personal number to another personal, terminal, or directory number when the PCS subscriber fails to answer or doesn't respond to paging. With personal call forwarding—no answer activated, a call coming in to the PCS subscriber will be automatically forwarded to the designated forward-to number whenever the PCS subscriber does not respond to the page or if the PCS subscriber does not answer within a specified period after transmission of the alert indication.

- **Call Forwarding—Unconditional** permits a PCS user to send incoming calls addressed to the PCS subscriber's personal number to another MS or directory number (forward-to number). The ability of the served PCS subscriber to originate calls is unaffected. If this service is activated, calls are forwarded independent of the state of the MS (busy, idle, etc.).

- **Call Transfer** permits a PCS user to transfer a call to another number on or off the PCS switch. When a call is transferred, the PCS personal terminal is then available for other calls.

- **Call Waiting** provides notification to a PCS subscriber of an incoming call while the user's mobile station is in the busy state. Subsequently, the user can either answer or ignore the incoming call. With call waiting activated, a new incoming call attempt to the PCS user who is already engaged in conversation on a prior call will receive a notification signal. This may be repeated a short time later if the PCS user takes no action. The calling party will hear an audible ringing signal either until the call attempt is aborted or the PCS user acknowledges the waiting call. The PCS user may indicate acceptance of the waiting call by (1) placing the existing call on hold or (2) releasing the existing call.

- **Calling Number Identification Presentation (CNIP)** is a supplementary service offered to a called party. It provides to the called party the number identification of the calling party. If the calling party has subscribed to calling number identification restriction, the calling number will not be presented.

- **Calling Number Identification Restriction (CNIR)** is a supplementary service offered to a calling party. It restricts presentation of that party's calling number identification to the called party. When the CNIR service is applicable and activated, the originating network provides the destination network with a notification that the calling number identification is not allowed to be presented to the called party. The CNIR may be offered with several options. Subscription options applied are (1) not subscribed (inactive for all calls); (2) permanently restricted (active for all calls); (3) temporarily restricted (specified by user per call)—default: restricted; and (4) temporarily allowed (specified by user per call)—default: allowed.

- **Conference Calling** is similar to three-way calling except when more than three parties are involved in the call.
- **Do Not Disturb** allows a wireless subscriber to direct that all incoming calls stop at the PCS switch and not page the mobile station.
- **Flexible Alerting** allows a call to a directory number to be branched into multiple attempts to alert several subscribers. The subscribers may have wireless or wireline terminations.
- **Message Waiting Notification** is the service where a message is sent to the MS to inform the user that there are messages stored in the network that the user can access.
- **Mobile Access Hunting** is the service where call delivery is presented to a series of terminating numbers. If the first number is not available, the system will try the second and continue down a list. The terminating numbers can be mobile or nonmobile numbers anywhere in the world.
- **Multilevel Precedence and Preemption (MLPP)** permits a group of wireless subscribers to have access to wireless service where higher-priority calls will be processed ahead of lower priority calls and may preempt (i.e., force the termination of) lower-level calls. Only calls within the same group will override each other.
- **Password Call Acceptance** is the service where calls to the wireless subscriber are interrupted and the calling party is asked to correctly enter a password before the mobile station is paged.
- **Preferred Language** is the capability for users to hear all network announcements in their preferred languages.
- **Priority Access and Channel Assignment** allows the service provider to provide capabilities to a subscriber that allows priority access to radio resources. This service permits emergency services personnel (e.g., police, fire, and rescue squads) priority access to the system. Multiple levels of access may be defined.
- **Remote Feature Call** permits a wireless subscriber to call a special directory number (from a wireless or wireline phone) and, after correctly entering account code information and a PIN, change the operation of one or more features of the service. For example, the selective call list can be modified by this capability.

- **Selective Call Acceptance** is the service where a wireless subscriber can form a list of those directory numbers that will result in the mobile station being paged. All other directory numbers will be blocked.
- **Subscriber PIN Access** is the ability to block access to the mobile station until the correct personal identification number (PIN) is entered into the MS.
- **Subscriber PIN Intercept** is the ability for a wireless subscriber to bar outgoing calls unless the correct PIN is entered. This feature can be implemented in the network or in the MS.
- **Three-Way Calling** permits a PCS user authorized for three-way calling to add a third party to an established two-way call regardless of which party originated the established call. To add a third party, the PCS user sends a request for three-way calling service to the service provider, which puts the first party on hold. The PCS user then proceeds to establish a call to the third party. A request by the controlling user, for disconnection of third party (i.e., last added party), will release that party and will cause the three-way connection to be disconnected and return the call to its original two-way state. If either of the noncontrolling parties to an established three-way call disconnects, the remaining two parties are connected as a normal two-way call. If the controlling PCS user disconnects, all connections are released.
- **Voice Message Retrieval** is service where the user can retrieve voice messages stored in the network. These messages are typically left by parties calling the user while the user was busy, did not answer, or was not registered with a system.
- **Voice Privacy** is the service where the user's voice traffic over the radio link is encrypted to prevent casual eavesdropping. With a personal communications system, in the United States, this is a required feature and is not optional.
- **Short Message Service** permits alpha and alphanumeric short messages to be sent to or from a mobile station.

4.4 SUMMARY

In this chapter, we presented the TR-45/TR-46 reference model, which is used by standards committees as the basis of describing network interfaces. The main elements of this model are the mobile station, base station, mobile switching center, home location register, and visited location register. Next, we discuss the MSC-BS interface (TIA IS-634), which standardizes the messaging between the base station and the mobile switching center. Messages between the BS and the MSC are categorized into two types: base station application part and direct transfer application part. Messages can be associated with one of the following functions: call processing and supplementary services, radio resource management, mobility management, and terrestrial facility management. The effects upon the architecture of a CDMA system are emphasized. Finally, we present a discussion about basic and supplementary services that are supported by cellular and PCS standards. From the point of view of the mobile subscriber, these services are independent of the air interface type (e.g., analog, TDMA, CDMA).

4.5 REFERENCES

1. TIA IS-634, "MSC-BS Interface for Public 800 MHz," Revision 0, 1995.
2. Committee T1—Telecommunications, "A Technical Report on Network Capabilities, Architectures, and Interfaces for Personal Communications," T1 Technical Report #34, May 1994.
3. Committee T1, "Stage 2 Service Description for Circuit Mode Switched Bearer Services," Draft T1.704.
4. TIA Interim Standard, IS-41 C, "Cellular Radio Telecommunications Intersystem Operations," January 1996.
5. TIA TR-45 Reference Model.
6. TIA TR-46 Reference Model.
7. TIA SP-2977, "Cellular Features Description," Prepublication version, March 14, 1995.
8. American National Standards Institute, Inc., "Synchronization Interface Standards for Digital Networks," 1087, ANSI T1.101-1987.
9. American National Standards Institute, Inc., "Signaling System No. 7 (SS7)—Message Transfer Part (MTP)," June 1992, ANSI T1.111-1992.
10. American National Standards Institute, Inc., "Signaling System No. 7 (SS7)—Signaling Connection Control Part (SCCP), June 1992, ANSI T1.112-1992.

Physical Layer of CDMA

5.1 INTRODUCTION

In chapter 2, we introduced the CDMA system used for cellular and personal communications systems in North America as embodied in the design standardized by the TIA and ATIS in IS-95A for a cellular system and J-STD-008 for a personal communications system. In this chapter, we will discuss this system in more detail and also describe the W-CDMA system standardized as J-STD-015 (and IS-665) for a personal communications system. The W-CDMA system is also expected to be used in Asia. The W-CDMA system is similar to the CDMA system, but there are some differences, which we will describe. In this chapter, we will discuss how a data (or voice-encoded) signal is modulated by a CDMA transmitter and demodulated by a CDMA receiver.

In CDMA, the entire 1.25-MHz transmission bandwidth is occupied by every station (see section 5.3 for the frequency bands used by a base station). In the forward direction (base station to mobile station), Walsh codes are used to distinguish different channels. On the reverse channel (mobile station to base station), different pseudorandom noise (PN) sequences are used to distinguish different channels. The Walsh functions are chosen so that the set of functions are all orthogonal to each other. All base stations in the system are on the same frequency and use the same set of time-shifted Walsh functions. Every base station in the system is synchronized to every other base station in the system. Different base stations use time-shifted versions of the PN sequence to permit mobile stations to select transmissions from different base stations. Thus, for CDMA, the frequency reuse factor N is 1. The PN sequences used by the MS are found by computer simulation and are chosen to have low autocorrelation and cross-correlation properties. The W-CDMA system uses Walsh functions in both directions for bandwidths of 5.0 and 10.0 MHz and Hadamard functions in both direction for a bandwidth of 15 MHz.

In this chapter, we first describe the Open System Interconnect reference model, which is used to construct the protocols for the CDMA and W-CDMA systems. We then examine the physical layer of the CDMA and W-CDMA systems in the rest of this chapter. In chapter 6, we examine the layer 2 and 3 functions for both systems; in chapter 7, we examine the applications functions (call processing); and in chapter 8, we examine the speech-coding methods.

5.2 OPEN SYSTEMS INTERCONNECT REFERENCE MODEL

Recently, the International Standards Organization (ISO) has developed a reference model for data communications. The Open Systems Interconnect (OSI) reference model is used by many computer systems for computer-to-computer communications. In the model, seven layers are defined to segment different aspects and needs for communications from each other so that the communications can be conducted in an orderly manner. In this section, we describe the OSI model in sufficient detail so that the protocols and messages in this chapter and in chapters 6 and 7 can be understood.

Each of the seven layers in the protocol communicates with its peer layer at the distant end and with the local layers immediately above and below it. The protocols at each layer define how the peer-to-peer communications takes place by defining message sets and state diagrams. For example, the layer 5 software in one computer communicates with the corresponding layer 5 software in another computer. The software in the two computers might be implemented in two different languages with two different operating systems and two completely different host computers. Thus, the layer 5 software that operates on one computer will not necessarily operate on another computer. But, as long as both layer 5 software packages agree on how they will meet the OSI specification, they will be able to communicate with each other. The model permits computer systems from widely different manufacturers to communicate with each other. For CDMA phones, the messaging is done within the OSI model and allows phones from any manufacturer to communicate with networks from any other manufacturer. In this chapter and chapters 6 and 7, we examine the details of the CDMA and W-CDMA systems and how they use the OSI model.

Figure 5.1 shows the OSI reference model for communications between computer systems. It shows two protocol stacks. One stack is for signaling and the other stack is for voice or data communications. With the reference model, each layer communicates with the layer immediately above and below it and with its peer layer at the other computer. With a properly designed system, the software at one layer can be replaced without affecting the other layers. Similarly, the software/hardware at the physical layer can be replaced without affecting the layers above it. In the following sections, we discuss each layer in more detail.

The seven layers of the model follow:

- **Layer 1. Physical Layer.** This layer describes the voltages or waveforms for a bit (1 and 0), the time duration of a bit, the pin connections and type of connector for baseband systems or the frequencies used for radio systems, the handshaking to start and stop a connection, and whether the connection is one way or two way.

- **Layer 2. Link Layer (Data Link).** This layer converts bits into frames of data. Methods must be determined for obtaining bit sync and frame sync of the frames, preventing the data from causing the false transmission of frame sync, and retransmitting data when errors occur or when frames are lost or duplicated. Buffers must be designed to cope with fast and slow transmitters and receivers.

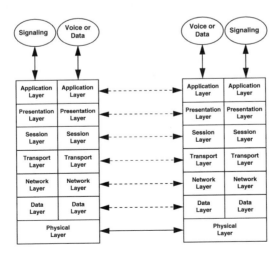

Figure 5.1 ISO Open Systems Interconnection reference model.

- **Layer 3. Network Layer.** This layer passes packets of information between two different end points. This layer often is designed to let layer 4 see an error-free channel. Unfortunately, this is not usually true. Thus, layer 4 must often also cope with errors. At this layer, billing and routing information for packets must be done. Buffering must be done at this layer to prevent too many packets from congesting the network.

- **Layer 4. Transport Layer.** The transport layer is the last layer to do error correction. All higher layers assume that the layer below it provides a perfect connection. Thus, the goal of this layer is to provide an error-free channel for the higher layers. If layer 3 is error-free, this job is easy. If layer 3 has errors, the transport protocol must allow for retransmission of data, error detection, and correction. This layer will set up and tear down calls to another host (addressing information is needed). It will also multiplex data from multiple processes in the host. Like other layers, it too must buffer data.

- **Layer 5. Session Layer.** This layer allows user or presentation layer processes to communicate. Events that occur during the layer are log-on messages and log-on IDs and passwords. Exchange of communications parameters occurs here (e.g., baud rate and full/half duplex). Grouping of messages can occur here. Automatic requests for a new connection can also occur here.

- **Layer 6. Presentation Layer.** This layer copes with things like protocol conversion, terminal type, encryption, definition of primitives, message compression, and file format conversion. Layer 6 software is sometimes combined with the layer 7 software, and sometimes the work done by the application layer is minimal and a null application layer is implemented.

- **Layer 7. Application Layer.** Here the definition of what goes on is up to the end user. Examples are forms entry, record locking, multiple hosts, and data base design.

- **Application.** Even though it is not normally shown in the OSI model protocol flow, the application software is above layer 7. For wireless phones, the application can be voice or data transmission for the user of the phone. The other main application is telephony functions (e.g., the mobility management function for a wireless phone).

In telephony, the signaling system application communicates directly with layer 3, and layers 4–7 are empty. Similarly, the traffic application (i.e., voice or data communications) communicates directly with layer 1, and layers 2–7 are empty. Figure 5.2 shows the CDMA system protocol stack mapped into the OSI reference model.

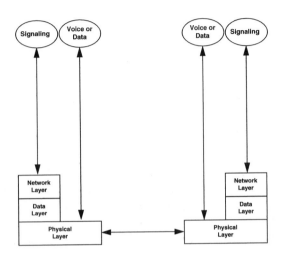

Figure 5.2 CDMA system mapped into the OSI reference model.

5.3 FORWARD CDMA CHANNEL AND W-CDMA CHANNEL

The forward CDMA channel consists of a pilot channel, an optional sync channel, optional (to a maximum of seven) paging channels, and several forward traffic channels. Each of these channels is orthogonally spread by the appropriate orthogonal function and is then spread by a quadrature pair of PN sequences. All the channels are added together and sent to the modulator. Many of the processes for constructing the forward and reverse channels are the similar; therefore, the details will be covered in section 5.6.

When a base station supports multiple forward CDMA channels, frequency division multiplex is used.

5.3.1 Pilot Channel

A pilot signal (fig. 5.3) is sent from the base station to aid in clock recovery at the MS. The pilot signal consists of the all zeros pattern and is modulo-2 added to the Walsh 0 function for the CDMA system. The W-CDMA system uses either Walsh 0 or Hadamard 0 but can use other Walsh codes as described in figure 5.3. The pilot signal is then sent to the modulator.

Figure 5.3 Forward CDMA pilot channel (all bandwidths).

5.3.2 Sync Channel

The sync channel is transmitted by a base station to enable the MS to obtain frame synchronization of the CDMA signal. The CDMA system uses a data rate of 1.2 kbps for the sync channel and then convolutionally encodes the data with a rate one-half code (see section 5.6). After encoding, the signal is processed by a symbol repetition stage and a block interleaver stage. The CDMA system repeats the symbol and then interleaves the data. The exact order of these two stages does not matter as long as the appropriate bits are in the correct place after the two stages. After interleaving, the resultant signal is modulo-2 added with the appropriate orthogonal code (see fig. 5.4) and then sent to the modulator.

The W-CDMA system uses a 16-kbps sync rate for all three bandwidths and then convolutionally encodes the data with a rate one-half code (see section 5.6). The W-CDMA system processes the output of the convolutional encoder as two separate data streams. After encoding, the signal is processed by a block interleaver stage and a symbol repetition stage. The exact order of these two stages does not matter as long as the appropriate bits are in the correct place after the two stages. After repetition, the resultant signal is modulo-2 added with the appropriate orthogonal code (see fig. 5.5) and then sent to the modulator.

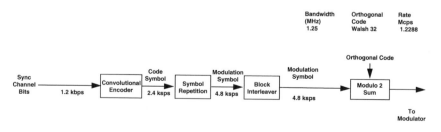

Figure 5.4 Forward CDMA sync channel.

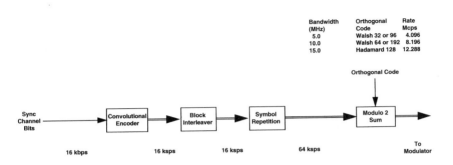

Figure 5.5 Forward W-CDMA sync channel.

5.3.3 Forward Paging Channel

The paging channel is transmitted by a base station to enable the MS to be paged and to process other orders while the MS is powered on and is idle. The CDMA system uses a data rate of either 4.8 or 9.6 kbps for the paging channel (fig. 5.6), and the W-CDMA system uses 16 kbps for the paging channel (fig. 5.7). Several of the modulation steps are similar to the sync channel. Both systems convolutionally encode the data with a rate one-half code (see section 5.6). After encoding, the signal is processed by a symbol repetition stage and a block interleaver stage. The encoding, repetition, and interleaving stages are identical to those used for the sync channel (and the traffic channel). For the CDMA system, the output modulation symbol is passed through a data scrambler. The scrambler prevents long sequences of 0s or 1s from appearing in the data stream. The scrambler is constructed using every 64th bit from a long code generator (see section 5.6) and modulo-2 summing it with the modu-

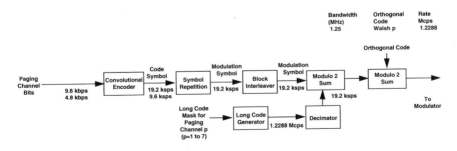

Figure 5.6 Forward CDMA paging channel.

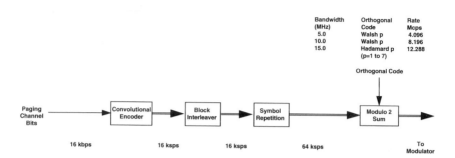

Figure 5.7 Forward W-CDMA paging channel.

lation symbol. The use of 1 out of 64 bits is called *decimation*. The W-CDMA system does not use a data scrambler. Finally, the resultant signal is modulo-2 added with the Walsh (or Hadamard) code. There are from 0 to 7 paging channels using codes 1 to 7, respectively and in sequence. Thus, for a base station with five paging channels, codes 1, 2, 3, 4, and 5 are used. The paging signal is then sent to the modulator.

5.3.4 Forward Traffic Channel

The traffic channel is transmitted by a base station to the MS (figs. 5.8 and 5.9) to carry voice or data traffic. The traffic channel is multiplexed and can carry voice or data, power control bits, and signaling channel data. The CDMA system multiplexes the voice, data, and signaling before the convolutional encoder. For CDMA, the data bits can be voice, data, or signaling and are multiplexed together according to the capabilities described in table 5.1. Signaling can be sent only by reducing the number of bits used for voice or data.

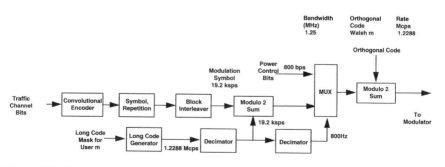

Figure 5.8 Forward CDMA traffic channel.

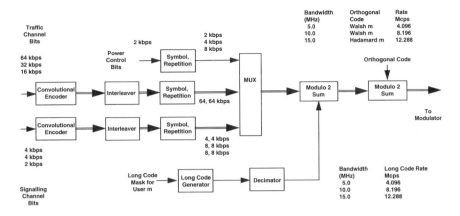

Figure 5.9 Forward W-CDMA traffic channel.

Table 5.1 Multiplexing Options for Forward CDMA Traffic Channel

Transmit Rate	Primary Traffic (voice or data), bits/frame	Signaling Traffic, bits/frame	Secondary Traffic (data), bits/frame
9600	171	0	0
	80	88	0
	40	128	0
	16	152	0
	0	168	0
	80	0	88
	40	0	128
	16	0	152
	0	0	168
4800	80	0	0
2400	40	0	0
1200	16	0	0

The W-CDMA multiplexes the channel after the symbol repetition and always processes a separate signaling channel. The CDMA system uses a data rate of 9.6 kbps for the traffic channel and the W-CDMA system uses 16, 32, or 64 kbps, depending on the type of voice encoding chosen. Both systems convolutionally encode the data with a rate one-half code (see section 5.6), repeat the symbols, and interleave the data using the same methods as the sync and paging channels. As is done on the paging channel, the CDMA system scrambles the data using the decimated long code. The power control bits are then multiplexed into the data stream. The resultant multiplexed signal is then further scrambled by the decimated long code, modulo-2 added to the Walsh (or Hadamard) code for the channel being used, and sent to the modulator. The long code mask chosen for each channel establishes the voice (or data) privacy for that channel.

5.3.5 Modulator

For the forward channel, the modulator (figure 5.10) is identical for both the CDMA and W-CDMA system. In the next section, we show that the reverse channel modulators are different. The I and Q signals from each channel (pilot, sync, paging, and traffic) are modulo-2 added to an I and Q pseudorandom noise sequence (see section 5.7.8 for the PN codes). For the CDMA system, the I and Q signals are identical, but the I and Q PN sequences are different. For the W-CDMA system, the I and Q signals are different and are derived from the processed data from the convolutionally encoded data; the same PN sequence is used for both the I and Q channels. The I and Q spread signals are then baseband filtered and the signals from all channels are sent to a linear adder with gain control. The gain control permits the individual channels to have different power levels assigned to them. The CDMA system assigns power levels to different channels depending on the quality of the received signal at a mobile sta-

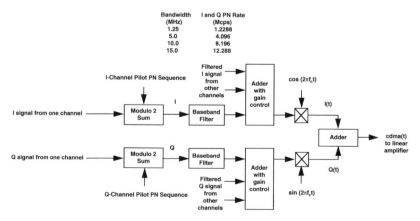

Figure 5.10 Forward CDMA modulator (all bandwidths).

tion. The algorithms for determining the power levels are proprietary to each equipment provider. The I and Q baseband signals are then modulated by the I and Q carrier signals, combined together, amplified, and sent to the antenna. The net signal from the CDMA modulator is a quadrature phase shift-signal [5, 6].

The same PN sequence is used on all channels (pilot, sync, paging, and traffic) of the CDMA forward channel. All base stations in a system are synchronized using the global positioning system satellites. Different base stations use time-shifted versions of the PN sequence to permit mobile stations to select the appropriate base station.

5.4 THE REVERSE CDMA CHANNEL

The reverse path from mobile station to base station uses a different frequency band. The reverse CDMA channel is composed of access channels and reverse traffic channels. All mobile stations accessing a base station over an access channel or a traffic channel share the same CDMA frequency assignment using direct sequence CDMA techniques. For CDMA, each traffic channel is identified by a distinct user long code sequence; each access channel is identified by a distinct access channel long code sequence. For W-CDMA, the channel selection is performed using Walsh (or Hadamard) codes similar to the forward channel. Multiple reverse CDMA channels may be used by a base station in a frequency division multiplexed manner.

The designers of the CDMA system assumed that recovery of the pilot signal from the mobiles would be difficult, so an asymmetric channel modulation method is used. On the reverse channel, Walsh functions are not used, but PN functions are used to distinguish the signals from different mobile transmitters. The designers of the W-CDMA system send a pilot signal and believe that they can recover the pilot signal. Therefore, W-CDMA systems uses Walsh (or Hadamard) functions in both directions.

5.4.1 Reverse Access Channel

The reverse access channel (figs. 5.11 and 5.12) is used by the MS to access the CDMA system to respond to pages, make call originations, and process other messages between the MS and the base station. The channel operates at 4.8 kbps for CDMA and 16 kbps for W-CDMA. The information bits are convolutionally encoded (rate 1/3 for CDMA and rate 1/2 for W-CDMA) and processed by symbol repetition and interleav-

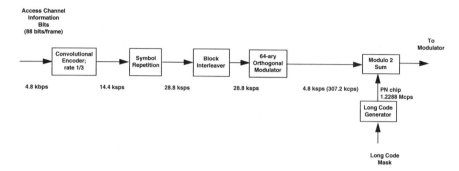

Figure 5.11 Reverse CDMA access channel.

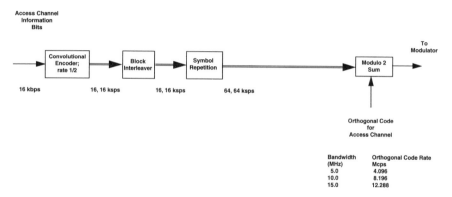

Figure 5.12 Reverse W-CDMA access channel.

ing functions. As in the forward channel, the order of the repetition and
interleaving stages is different for CDMA and W-CDMA. In practice,
they could be done in either order as long as the bits are in the correct
position after the two stages. The W-CDMA also separately processes the
output from each convolutional encoder (as it did on the forward channel). The CDMA and W-CDMA systems operate differently in forming
the orthogonal modulation spreading. The CDMA system processes each
code symbol through an orthogonal 64-ary modulator generating a
Walsh symbol at rate 307.2 kilochips per second (kcps) for each input
symbol. It then modulo-2 sums the signal with a PN long code. The long
code is chosen by computer simulations to have good orthogonal properties so that the base station can select the MS transmissions. In the W-
CDMA system, output of the symbol repetition stage is modulo-2 added
to a Walsh (or Hadamard) code for the access channel. The orthogonal

properties of the Walsh (or Hadamard) codes is used by the base station to select MS transmissions. The orthogonally spread signal is then sent to the modulator.

5.4.2 Reverse Traffic Channel

For the CDMA system, the primary traffic channel, the secondary traffic channel, and the signaling channel (fig. 5.13) are multiplexed together (see table 5.1 for multiplexing options) and processed by the same covolutional encoder, symbol repetition, interleaver, and 64-ary orthogonal modulator that is used for the access channel. The modulator output is then randomized (to eliminate repetitive 0s and 1s patterns) by blocks of 14 bits taken from the long code. The output of the randomizer is then spread by the long code. The orthogonally spread signal is then sent to the modulator.

For the W-CDMA system, the channel multiplexing is done in the modulator. The output of the traffic channel (encoded speech at 16, 32, or 64 kbps or data at rates up to 64 kbps) is convolutionally encoded, block interleaved, and repeated (if the data rate is less than 64 kbps). The output signal (fig. 5.14) is then modulo-2 summed with the orthogonal code (Walsh or Hadamard) for the traffic channel and sent to the modulator.

For the W-CDMA system, signaling occurs on the traffic channel at either 2 or 4 kbps. The output of the signaling channel (fig. 5.15) is convolutionally encoded, block interleaved, and repeated to generate a symbol rate of 64 kilosymbols per second (ksps). The output signal is then modulo-2 summed with the orthogonal code (Walsh or Hadamard) for the signaling channel and sent to the modulator.

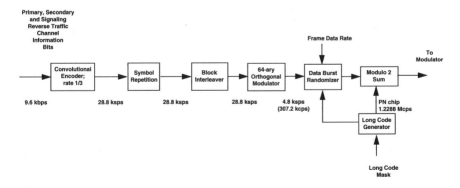

Figure 5.13 Reverse CDMA traffic channel.

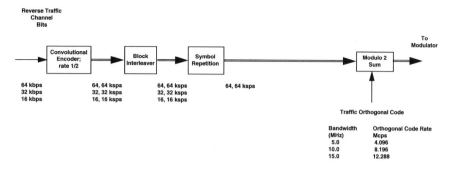

Figure 5.14 Reverse W-CDMA traffic channel.

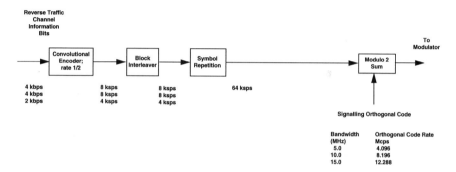

Figure 5.15 Reverse W-CDMA traffic information channel.

5.4.3 Reverse Channel Modulator

Unlike the forward direction, the CDMA and W-CDMA systems use different modulation methods to generate the CDMA signal. The net signal from either CDMA modulator is a four-phase quadrature signal, but the characteristics are different.

For the CDMA system, the output from either the access channel or the traffic channel is sent to two modulo-2 adders, one for the in-phase and one for the quadrature channel (fig. 5.16). Two different PN sequences are modulo-2 added to the data and filtered by a baseband filter. For the quadrature channel, a delay of 1/2 of a PN symbol (406.9 ns) is added before the filter. Thus, the reverse channel uses offset quadrature phase-shift keying (O-QPSK). No pilot signal is transmitted on the reverse channel.

For the W-CDMA system, the in-phase and quadrature signals were separated after convolutional encoding (fig. 5.17). The I channel linearly

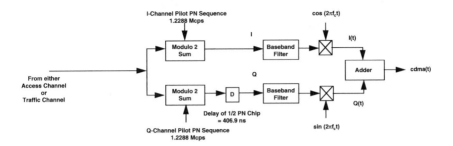

Figure 5.16 Reverse CDMA modulator.

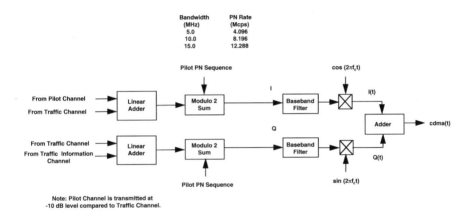

Figure 5.17 Reverse W-CDMA modulator.

adds the traffic channel data and a reduced level pilot channel (−10 dB below the traffic channel). The Q channel linearly adds the other traffic channel and the traffic information channel. Both the I and the Q channels are then modulo-2 summed with the same pilot PN sequence, bandpass filtered, and sent to the modulator. No phase delay is used on the Q channel, so the resultant modulation is quadrature phase-shift keying.

5.5 CHANNEL SPACING AND FREQUENCY TOLERANCE

CDMA uses a bandwidth of 1.25 MHz and defines a set of channels on a 50-kHz spacing for J-STD-008 and 30-kHz spacing for IS-95A. Within the assigned cellular and PCS spectrum, any frequencies can be used as long as the signal remains within the assigned spectrum for a service provider. Since mobile stations must know where to listen for base station transmissions, there is a preferred set of channels for use with CDMA. Otherwise, mobile stations would need to enter a lengthy search process to find a CDMA channel. Once a channel is assigned to CDMA (preferred on nonpreferred), additional CDMA channels must remain on a 1.25-MHz channel spacing.

For cellular frequencies, the entire cellular frequency band is segmented into an A service provider and a B service provider band with subbands. The subbands were added when the FCC allocated additional channels to cellular telephones. The cellular channels for CDMA are described in tables 5.2 and 5.3. Certain channels are not permitted and are declared not valid since use of those channels would cause the CDMA signal to be outside of the assigned band.

Table 5.2 Definition of Valid Channel Numbers for CDMA at Cellular Frequencies.
(Reproduced under written permission of the copyright holder [TIA].)

Frequency Band	Bandwidth of Band, MHz	Valid Regions	Channel Number
A"	1	Not valid	991–1012
		Valid	1013–1023
A	10	Valid	1–311
		Not valid	312–333
A'	1.5	Not valid	667–688
		Valid	689–694
		Not valid	695–716
B	10	Not valid	334–355
		Valid	356–644
		Not valid	645–666
B'	2.5	Not valid	717–738
		Valid	739–777
		Not valid	778–799

Table 5.3 Definition of Preferred Channel Numbers for CDMA at Cellular Frequencies.
(Reproduced under written permission of the copyright holder [TIA].)

Band	Preferred CDMA Channel Number	MS Transmit Center Frequency, MHz	Base Station Transmit Center Frequency, MHz
A	283	833.490	878.490
A'	691	845.730	890.730
A"	None: band is less than 1.25 MHz	—	—
B	384	836.520	881.520
B'	777	848.310	893.310

At PCS frequencies, six band segments are defined, thus permitting up to six service providers in each area. Tables 5.4 and 5.5 define the permitted and preferred channel numbers for CDMA at PCS frequencies.

Table 5.4 Definition of Channel Numbers for CDMA at PCS Frequencies. (Reproduced under written permission of the copyright holder [TIA].)

Band	Valid CDMA Frequency Assignments	CDMA Channel Number	MS Transmit Center Frequency, MHz	Base Station Transmit Center Frequency, MHz
A	Not valid	0–24	1850.000–1851.200	1930.000–1931.200
(15 MHz)	Valid	25–275	1851.250–1863.750	1931.250–1933.750
	Cond. valid	276–299	1863.800–1864.950	1933.800–1934.950
D	Cond. valid	300–324	1865.000–1866.200	1945.000–1946.200
(5 MHz)	Valid	325–375	1866.250–1883.750	1946.250–1943.750
	Cond. valid	376–399	1868.800–1869.950	1948.800–1949.950
B	Cond. valid	400–424	1870.000–1871.200	1950.000–1951.200
(15 MHz)	Valid	425–675	1871.250–1883.750	1951.250–1963.750
	Cond. valid	676–699	1883.800–1884.950	1963.800–1964.950
E	Cond. valid	700–724	1885.000–1886.200	1965.000–1966.200
(5 MHz)	Valid	725–775	1886.250–1883.750	1966.250–1963.750
	Cond. valid	776–799	1888.800–1889.950	1968.800–1969.950
F	Cond. valid	800–824	1890.000–1891.200	1970.000–1971.200
(5 MHz)	Valid cond	825–875	1891.250–1893.750	1971.250–1973.750
	Valid	876–899	1893.800–1894.950	1973.800–1974.950
C	Cond. valid	900–924	1895.000–1896.200	1975.000–1976.200
(15 MHz)	Valid	935–1175	1896.250–1908.750	1976.250–1988.750
	Not valid	1176–1199	1908.800–1909.950	1988.800–1989.950

Table 5.5 Definition of Preferred Channel Numbers for CDMA at PCS Frequencies. (Reproduced under written permission of the copyright holder [TIA].)

Block Designator	Preferred Set of Channel Numbers
A	25, 50, 75, 100, 125, 150, 175, 200, 225, 250, 275
D	325, 350, 375
B	425, 450, 475, 500, 525, 550, 575, 600, 625, 650, 675
E	725, 750, 775
F	825, 850, 875
C	925, 950, 975, 1000, 1025, 1050, 1075, 1100, 1125, 1150, 1175

The base station is the master reference for the CDMA system, and it must maintain its frequency to within ±5 parts per 100 million (±100 Hz at 2000 MHz). The MS must maintain its transmit carrier frequency 80 MHz ± 150 Hz below the base station frequency for PCS frequencies and 45 MHz ± 300 Hz below the base station frequency for cellular frequencies.

For W-CDMA, a 5-MHz band will use a single 5-MHz W-CDMA system. A 15-MHz spectrum will use either one 15-MHz system, one 10-MHz and one 5-MHz system, or three 5-MHz systems.

The W-CDMA base station transmitter must maintain its frequency to within ± 5 parts per 100 million (± 100 Hz at 2000 MHz). The MS transmit carrier frequency will be 80 MHz (± 200 Hz) below the corresponding base station transmit signal.

5.6 POWER CONTROL IN CDMA

The CDMA system defines three power classes at cellular frequencies and five power classes at PCS frequencies (see tables 5.6 and 5.7) for MSs. The MS attempts to control the power output based on received signal strength (open loop control), and the base station sends power control messages to the MS about once every millisecond (closed loop control). The net effect is to control the power received at the base station to within 1 dB for all MSs being received at that base station. The fine level of power control is necessary for proper operation of the CDMA system.

The base station can also vary its transmitted power by ± 4 dB depending on the error rates reported by the mobile station [5]. The mobile station can report frame error rates (as measured on the forward traffic channel) by sending either a power measurement report message (either rate set) or by setting the erasure indicator bit (for rate set 2 only) as described in TSB-74 [7]. Details on how the base station uses reported error rates to control power is proprietary to an equipment vendor's design.

Table 5.6 Maximum Effective Isotropic Radiated Power for a Cellular CDMA MS. (Reproduced under written permission of the copyright holder [TIA].)

Mobile Station Class	EIRP at Maximum Output Shall Exceed	EIRP at Maximum Output Shall Not Exceed
I	+1 dBW (1.25 W)	+8 dBW (6.3 W)
II	–3 dBW (0.5 W)	+4 dBW (2.5 W)
III	–7 dBW (0.2 W)	0 dBW (1.0 W)

Table 5.7 Maximum Effective Isotropic Radiated Power for a PCS CDMA MS. (Reproduced under written permission of the copyright holder [TIA].)

Mobile Station Class	EIRP at Maximum Output Shall Exceed	EIRP at Maximum Output Shall Not Exceed
I	–2 dBW (0.63 W)	3 dBW (2.0 W)
II	–7 dBW (0.20 W)	0 dBW (1.0 W)
III	–12 dBW (63 mW)	–3 dBW (0.5 W)
IV	–17 dBW (20 mW)	–6 dBW (0.25 W)
V	–22 dBW (6.3 mW)	–9 dBW (0.13 W)

The W-CDMA system defines three power classes of PCS mobile stations (see table 5.8).

Table 5.8 Maximum Effective Isotropic Radiated Power for W-CDMA MS. (Reproduced under written permission of the copyright holder [TIA].)

Mobile Station Class	EIRP at Maximum Output Shall Exceed
I	23 dBm (200 mW)
II	13 dBm (20 mW)
III	3 dBm (2 mW)

The CDMA mobile station gates its power on and off depending on the data to be transmitted. This feature is useful for voice-encoded data to improve system performance since the transmitter will not be on during gaps in speech. When the transmitter is gated off, it must reduce its power by 20 dB. If the 20-dB reduction would be below the noise floor of the transmitter, it is permitted to gate the transmitter off to the power level of the noise floor of the transmitter.

When an MS attempts an access on the reverse channel, it must transmit at a power level of

$$P = P_{mean} + \text{NOM\_PWR} + \text{INT\_PWR} - \text{P\_CNST dBm} \qquad (5.1)$$

where P_{mean} = the mean input power of the MS transmitter;
NOM_PWR = the nominal correction factor for the base station, as defined in the overhead message;
INT_PWR = the correction factor for the base station from partial path loss decorrelation between transmit and receive frequencies, as defined in the overhead message;
P_CNST = 73, a constant.

If the access is not successful, the MS will increase its power by PWR_STEP (as defined on the overhead message), remember the total number of unsuccessful accesses it has made (and the sum of all corrections called access probe corrections), and try again, if necessary. It will continue, until it is successful or the process is stopped by the access attempt procedures (see chapter 7). If the maximum power is reached, the MS maintains that maximum power.

When the MS transmits on the reverse traffic channel, it uses a power of

$$P = P_{mean} + \text{NOM\_PWR} + \text{INT\_PWR} - \text{P\_CNST}$$
$$+ \text{ sum of all access probe corrections dBm} \qquad (5.2)$$

Once communication with the base station occurs, the base station will measure the received signal strength and send closed loop power control messages. Then the power output of the MS will be

$$P = P_{mean} + \text{NOM\_PWR} + \text{INT\_PWR} - \text{P\_CNST}$$
$$+ \text{ sum of all access probe corrections}$$
$$+ \text{ sum of all closed loop power control corrections dBm} \qquad (5.3)$$

The ranges and typical values for the power control parameters are

$$-8 < \text{NOM\_PWR} < 7 \text{ dB} \qquad (5.4)$$

$$\text{Typical NOM\_PWR} = 0 \text{ dB} \qquad (5.5)$$

$$-16 < \text{INIT\_PWR} < 15 \text{ dB} \qquad (5.6)$$

$$\text{Typical INIT\_PWR} = 0 \text{ dB} \qquad (5.7)$$

$$0 < \text{PWR\_STEP} < 7 \text{ dB} \qquad (5.8)$$

The values of these parameters for each base station are transmitted on the forward channel in the Access Parameters message (see chapter 7).

The base station transmits separate power control bits for each of the MS transmitting on the reverse channel. When an MS receives a power control bit, it will increase or decrease its power by 1 dB (bit value of 0 = increase; bit value of 1 = decrease). The MS will maintain a cumulative sum of all received power control bits to determine the correct power output to use. The total range of power control is ±24 dB around the open loop estimated power.

The base station power is limited to 1640 W of effective isotropic radiated power (EIRP) in any direction in a 1.25-MHz band for antenna heights above average terrain (HAAT) less than 300 m. When the base station antenna height exceeds 300 m, the EIRP must be reduced according to current FCC rules.

For PCS frequencies, the base station power output in any direction should not exceed 100 W.

For the W-CDMA system,

$$P\_CNST = -61 \tag{5.9}$$

$$-47 < INIT\_PWR < 11 \text{ dB} \tag{5.10}$$

$$\text{Typical } INIT\_PWR = 0 \text{ dB} \tag{5.11}$$

No range for the nominal power (NOM_PWR) is specified. Its value is obtained from the forward overhead channel parameters.

5.7 MODULATION PARAMETERS

In sections 5.2 and 5.3, we discussed the various modulation stages and their order in processing the different bit steams on the forward and reverse channels. In this section, we will describe each of the modulation stages in more detail.

5.7.1 Convolutional Encoding

With convolutional encoding (see fig. 5.18), the encoded output is a function of the input bit stream and delayed versions of the bit stream:

$$c(t) = \sum_{n=0}^{N-1} a_n i(t - n\tau) \tag{5.12}$$

where $i(t)$ = information bit stream,
   $c(t)$ = output bit stream,
   a_n = coefficient (0 or 1) to specify the addition of the
         delayed version of $i(t)$,
   τ = bit symbol duration, and
   N = length of the code.

The coefficients a_n are often written as a generator function

$$g = \begin{bmatrix} a_0 \, a_1 \, ... \, a_{N-2} \, a_{N-1} \end{bmatrix} \tag{5.13}$$

or generator polynomial

$$g(x) = a_0 x^0 + a_1 x^1 + a_2 x^2 + ... + a_{N-1} x^{N-1} \tag{5.14}$$

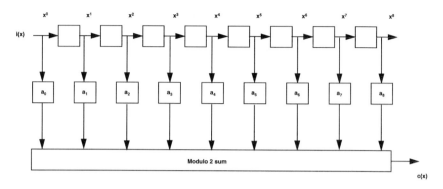

Figure 5.18 Block diagram of convolutional encoder.

For the CDMA MS, a rate one-third convolutional code is used, where three output bits are generated for each input bit. Thus, for an input signal $i(t)$, the output is

$$c_0(t) = g_0(t) \bullet i(t)$$
$$c_1(t) = g_1(t) \bullet i(t) \qquad (5.15)$$
$$c_2(t) = g_2(t) \bullet i(t)$$

where three bits c_0, c_1, and c_2 are transmitted in sequence, at three times the information rate, for each bit into the encoder.

The generator codes are

$$g_0 = \begin{bmatrix} 1 & 0 & 1 & 1 & 0 & 1 & 1 & 1 & 1 \end{bmatrix}$$
$$g_1 = \begin{bmatrix} 1 & 1 & 0 & 1 & 1 & 0 & 0 & 1 & 1 \end{bmatrix} \qquad (5.16)$$
$$g_2 = \begin{bmatrix} 1 & 1 & 1 & 0 & 0 & 1 & 0 & 0 & 1 \end{bmatrix}$$

For the base station, a rate one-half code is used, with generator codes of

$$g_0 = \begin{bmatrix} 1 & 1 & 1 & 1 & 0 & 1 & 0 & 1 & 1 \end{bmatrix}$$
$$g_1 = \begin{bmatrix} 1 & 0 & 1 & 1 & 1 & 0 & 0 & 0 & 1 \end{bmatrix} \qquad (5.17)$$

For the W-CDMA system, a rate one-half code is used in both directions with generator codes specified in equation (5.17). However, the output of the two bit streams is processed in parallel and carried through to further modulation stages as two parallel bit streams.

5.7.2 Bit Repetition

The nominal data rate on the forward and reverse CDMA channels is 9600 bps. If data are being transmitted at a lower rate (4800, 2400, or 1200 bps), then the data bits are repeated n times to increase the rate to 9600 bps.

For the W-CDMA system, the nominal data rate is 64 kbps. Therefore any channel-transmitting data at less than 64 kbps are rate multiplied up to a constant 64-kbps rate.

5.7.3 Block Interleaving

The communications over a radio channel are characterized by deep fades that can cause large numbers of consecutive errors. Most coding systems perform better on random data errors rather than on blocks of errors. By interleaving the data, no two adjacent bits are transmitted near to each other, and the data errors are randomized. The specifics of the interleaver are specified in the respective standards and are not reproduced here.

For CDMA, the interleaver spans a 20-ms frame. In the reverse direction, the output of the interleaver is a fixed 28.8 ksps. If the data rate is 9.6 kbps, the resultant signal transmits with a 100 percent duty cycle. If the data rate is lower (4800, 2400, or 1200 bps), the interleaver plus the randomizer (see the next section) deletes redundant bits and transmits with a lower duty cycle (50, 25, 12.5 percent). Thus, bits are not repeated on the reverse CDMA traffic channel. On the access channel, the data bits are repeated. In the forward direction, the nominal data rate is 19.2 kbps, and lower data rates use a lower duty cycle.

For the W-CDMA system (at all three bandwidths), the block interleaver span is 5 ms (the traffic channel spans of 10 and 20 ms are supported as options). Different interleaver matrices are used for different channels. All channels operate at a constant 64-kbps rate.

As noted in sections 5.2 and 5.3, the CDMA and W-CDMA systems differ in the order of repetition and interleaving.

5.7.4 Randomizing

For the CDMA system only, and only on the reverse traffic channel, the output of the interleaver is processed by a data randomizer. The randomizer removes redundant data blocks generated by the code repetition. It uses a masking pattern determined by the data rate and the last 14 bits of the long code. For each 20-ms block (192 bits at 9600 bps), the data randomizer segments the block into 16 blocks of 1.25 ms. At a data rate of 9600 bps, all blocks are filled with data. At a data rate of 4800 bps, 8 of the 16 blocks are filled with data in a random manner. Similarly, for 2400 and 1200 bps, 4 of the 16 and 2 of the 16 blocks, respectively, are randomly filled with data. Thus, no redundant data are sent over the reverse channel.

The W-CDMA system does not use a randomizer in either direction.

5.7.5 Orthogonal Codes

In the forward direction for both CDMA systems and in the reverse direction for the W-CDMA system, the data streams are modulo-2 added to an orthogonal code. The orthogonal code is 1 of 64 Walsh functions for CDMA and 1 of 256 Walsh functions for W-CDMA at 5-MHz bandwidth. At 10-MHz bandwidth, the code is 1 of 512 Walsh codes, and at 15-MHz bandwidth, the code is 1 of 768 Hadamard codes. All codes are orthogonal to each other (see chapter 2 for a more detailed discussion of the orthogonal codes).

For the reverse W-CDMA channel, the orthogonal code is the modulo-2 sum of the Walsh or Hadamard codes and the long code (see fig. 5.19).

5.7.6 64-ary Orthogonal Modulation

For the CDMA system only, and only on the reverse channel, the data stream is modulated by a 64-ary orthogonal modulator. For each six

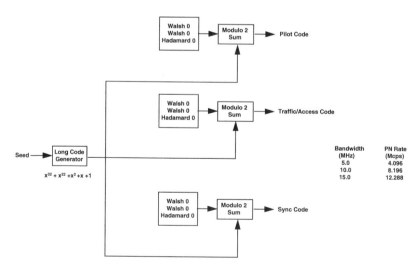

Figure 5.19 Orthogonal code generation for W-CDMA.

input symbols to the modulator, one output Walsh function is generated. The output Walsh function is defined by

$$W_i = c_0 + 2c_1 + 4c_2 + 8c_3 + 16c_4 + 32c_5 \qquad (5.18)$$

where c_5 is the most recent and c_0 the oldest of the 6 bits to be transmitted and W_i is chosen from 1 of 64 orthogonal Walsh functions. The W-CDMA system does not use this step.

5.7.7 Long Codes

Prior to the modulation stage, the reverse CDMA signal is spread by a long code at a rate of 1.2288 Mcps. The long code has a length of $2^{42} - 1$ bits and is generated by the following polynomial:

$$\begin{aligned} l(x) = \ & x^{42} + x^{35} + x^{31} + x^{27} + x^{26} + x^{25} + x^{22} + x^{21} + x^{19} + \\ & x^{18} + x^{17} + x^{16} + x^{10} + x^7 + x^6 + x^5 + x^3 + x^2 + x + 1 \end{aligned} \qquad (5.19)$$

The output of the long code generator is modulo-2 added with a 42-bit long code mask to generate the long code. The mask depends on the channel used (access or traffic) and the information on the MS transmitting the data. On the traffic channel, either a public or private long code mask is used. The public mask is a function of the electronic serial number of the MS. The private long code mask is generated by a secret algorithm used for voice and data privacy. On the access channel, the long code mask is generated by combinations of data associated with the access channel.

Time alignment of the long code is generated by defining time equal to zero to be January 6, 1980 at 00:00:00 UTC.

5.7.8 Direct PN Spreading

Both the CDMA and W-CDMA systems modulate the processed data stream with a pseudorandom noise spreading sequence at the fundamental rate for the system (1.2288, 4.096, 8.192, or 12.288 Mcps).

For the forward and reverse CDMA system, the following codes are used to generate the PN sequence:

$$g_I(x) = x^{15} + x^{13} + x^9 + x^8 + x^7 + x^5 + 1 \qquad (5.20)$$

$$g_Q(x) = x^{15} + x^{12} + x^{11} + x^{10} + x^6 + x^5 + x^4 + x^3 + 1 \qquad (5.21)$$

For the W-CDMA system, the following code is used for both the I and the Q channels:

$$g(x) = x^{32} + x^{22} + x^2 + x + 1 \qquad (5.22)$$

The PN sequence generator is seeded with data received in messages sent from the base station. The seed is used to establish the voice and data privacy on the channel. The same seed is used in both directions.

5.7.9 Baseband Filtering

After PN modulation, the signal is filtered by a baseband filter. The filter should have the following parameters:

Passband ripple: 3 dB

Upper passband frequency: 590 khz

Minimum stopband attentuation: 40 dB

Lower stopband frequency: 740 khz

For the W-CDMA system, the following values hold (see table 5.9).

Table 5.9 Baseband Filter Parameters for W-CDMA System

System Bandwidth, MHz	Passband Ripple, dB	Upper Passband Frequency, MHz	Minimum Stopband Attenuation, dB	Lower Stopband Frequency, MHz
5.0	3	1.96	40	2.47
10.0	3	3.92	40	4.94
15.0	3	5.88	40	7.41

5.7.10 Synchronization of CDMA Signals

Time 0 for the CDMA system is January 6, 1980 at 00:00:00 UTC. This is the same as time 0 for the global positioning system (GPS); therefore, CDMA time is the same as GPS time. GPS time and UTC time differ by the number of leap seconds since January 6, 1980. Thus, GPS time and UTC time are synchronous but can differ by an integer number of seconds.

All BSs in a CDMA system are synchronized to the GPS. Each base station transmits a set of orthogonal codes that are synchronized to CDMA time and are time-shifted from the codes at other BSs in the system.

5.8 SUMMARY

The International Standards Organization has developed the Open Systems Interconnect reference model for data communications that is used by most computer systems to design the communications protocols. In this chapter, we first reviewed the seven layers of the OSI model. We then limited discussion in this chapter to the physical layer (layer 1) of the CDMA and W-CDMA systems as defined in the ATIS and TIA standards. Both systems have the same goal: to efficiently use the available spectrum to provide digital cellular and PCS services. The CDMA system uses 64 orthogonal Walsh codes at a data rate of 1.2288 Mbps to code the digital signals for voice, data, and control. The W-CDMA system uses Walsh or Hadamard codes at higher bit rates (4.096, 8.192 and 12.288 Mbps) to accomplish the same result. The designers of the CDMA system do not attempt to recover pilot signal synchronization on the reverse channel (from mobile station to base station). They, therefore, use 64-ary modulation (with 1 of 64 Walsh symbols) on the reverse channel and use pseudorandom noise sequences to obtain the orthogonal modulation. The designers of the W-CDMA system believe that synchronization can be obtained on the reverse channel (at the higher data rates) and, therefore, use Walsh or Hadarmard codes in both directions. The two systems have other minor differences in the ordering between the various encoding steps but are otherwise similar.

Since the CDMA system can be placed anywhere in the cellular or PCS band, the standards define a set of preferred channels. A service provider can use any of the preferred channels. Because a critical component of CDMA systems is the ability to control the power of the mobile station almost instantaneously, we reviewed the power control methods used in both systems. We then concluded the chapter by defining the detailed modulation steps used in both systems and examined the differences between the two systems.

The CDMA system is defined by two standards: IS-95A for a dual-mode analog/digital system for cellular frequencies and J-STD-008 for a digital-only system at PCS frequencies. While there may be minor differences between IS-95A and J-STD-008 for the digital implementations, the goal of the two standards committees is that the digital part of both standards be identical except for frequency bands used. For the W-CDMA system, the standard is defined for PCS frequencies only, and one standard has two references numbers (IS-665 and J-STD-015). For more information on the specifics of the physical layer, consult the standards.

5.9 REFERENCES

1. TIA IS-95A, "Mobile Station–Base Station Compatibility Standard for Dual Mode Spread Spectrum Cellular System."

2. Alliance for Telecommunications Industry Standards J-STD-008, "Personal Station–Base Station Compatibility Requirements for 1.8 to 2.0 GHz Code Division Multiple Access (CDMA) Personal Communications Systems."

3. TIA IS-665, "W-CDMA (Wideband Code Division Multiple Access) Air Interface Compatibility Standard for 1.85-1.99 GHz PCS Applications."

4. Alliance for Telecommunications Industry Standards J-STD-015, "W-CDMA (Wideband Code Division Multiple Access) Air Interface Compatibility Standard for 1.85-1.99 GHz PCS Applications."

5. "Radio System Characterization for the Proposed IS-95 based CDMA PCS Standard," Joint Technical Committee (on Air Interface Standards) of T1P1 and TR-46 contribution JTC(Air)/94.11.03-735, November 3, 1994.

6. Whipple, D. P., "The CDMA Standard," *Applied Microwave and Wireless*, 6(2), Spring 1994, pp. 24–37.

7. TIA PN-3570 (TSB-74), "Telecommunications Systems Bulletin: Support for 14.4 kbps Data Rate and PCS Interaction for Wideband Spread Spectrum Cellular Systems," October 1995.

Network and Data Link
Layers of CDMA

6.1 INTRODUCTION

In this chapter, we explore layers 2 and 3, the data link layer and the
network layer used for CDMA, and the detailed messages that are sent
over the CDMA air link. Some of the messages are sent only between the
base station and the mobile station. Other messages are sent between
the MS and other network elements. In chapter 7, we use these messages
to show how call processing flows across the network. Information flows
from the BS to the MS via the forward CDMA channel, on the pilot channel, the sync channel, the paging channel, and the forward traffic channel. Information flows from the MS to the BS on either the access
channel or the reverse traffic channel. The BS/MS communications take
place on the paging/access channel during call setup and on the forward/
reverse traffic channel during a call.

All cellular and personal communications systems air interfaces
(except GSM) used in North America share a common approach to the
operation of a MS. In chapter 4, we discussed the high-level operation of
the MS as it implements the common operational approach. In chapter 7,
we discuss how the CDMA and W-CDMA systems use the operations
described there to provide services; and in chapter 8, we discuss the voice
coding systems used for CDMA.

Both CDMA and W-CDMA define control channels (sync, paging,
and access channels) that are used for data communications between the
MS and the PCS/cellular system and traffic channels that are used for
user-to-user communications (voice or data).

121

The CDMA system combines the operation of the network and data link layers and treats them as one layer. The W-CDMA system uses higher-speed signaling and voice-encoding rates and takes an approach that is similar to ISDN and, thus, treats layers 2 and 3 as separate and distinct layers.

In this chapter, we discuss the detailed message framing for both systems and describe some of the typical messages that are sent in the system. There are many services supported in the CDMA system, and we encourage you to consult the applicable standards for a full treatment of the many messages.

When an MS is first powered up, it must find and decode data on a control channel before any further processing can be done. For the messages described in this chapter, we assume that the BS to MS channels are properly synchronized in the receivers of both sides and that the receivers are properly decoding data. The operations necessary for these events to happen are classified as engineering art and are usually proprietary to a given manufacturer of equipment. Some manufacturers provide integrated circuit chip sets to perform the proper data modulation and demodulation. The encoding and decoding of the messages described in this chapter are typically performed in the software (or firmware) of the BS and MS.

6.2 FORWARD CDMA CHANNEL

Data can be transmitted from a BS to a MS over the sync channel, the paging channel, or the information stream on the forward traffic channel. Some of the data are specific to a particular channel. Other data (e.g., orders) can be sent on the paging channel or the traffic channel.

6.2.1 Sync Channel

The forward sync channel operates at a data rate of 1200 bps and transmits information that is specific to the BS and needed by the MS to access the system.

The Sync Channel message (fig. 6.1) has an 8-bit message length header, a message body of a minimum of 2 bits and a maximum of 1146 bits, and a cyclic redundancy check (CRC) code of 30 bits. If the sync channel messages are less than an integer multiple of 93 bits, they are padded with 0 bits at the end of the message. The message length includes the header, body, and CRC, but not the padding. The CRC is computed on the message length header and the message body using the following code:

$$g(x) = x^{30} + x^{29} + x^{21} + x^{20} + x^{15} + x^{13} + x^{12} + x^{11} + x^{8} + x^{7} + x^{6} + x^{2} + x + 1 \quad (6.1)$$

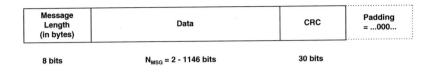

Message Length (in bytes)	Data	CRC	Padding = ...000...
8 bits	N_{MSG} = 2 - 1146 bits	30 bits	

Notes: N_{MSG} = Message length in bits (including length field and CRC)

Padding bits are not used for Unsynchronized Paging Channel Messages
Sync   Channel Data Rate = 1200 bps
Paging Channel Data Rate = 4800 bps or 9600 bps

Figure 6.1 CDMA message framing on forward sync channel and paging channel.

After a message is formed, it is segmented into 31-bit groups and sent in a sync channel frame (fig. 6.2) consisting of a 1-bit start of message (SOM) field and 31 bits of the sync channel frame body. A value of 1 for SOM indicates that the frame is the start of a Sync Channel message. A value of 0 for SOM indicates that the frame is a continuation of a Sync Channel message or padding.

Three sync channel frames are combined to form a sync channel superframe (fig. 6.3) of length 80 ms (96 bits). The entire sync channel message is then sent in N superframes. The padding bits are used so that the start message always starts at one bit after the beginning of a superframe. The first bit of the superframe is SOM = 1.

The only message sent on the sync channel is the Sync Channel message that transmits information about the BS and the serving CDMA system. Some of the information being sent follows.

One set of data, the system identification (SID) and the network identification (NID), define the system being received and the network within the system. The values for SID and NID are defined by the Federal Communications Commission.

Other data define the offset of the PN sequence for the BS and the long code state for that BS.

The sync channel also sends information about the system time, leap seconds, offset from UTC, and the state of daylight savings time. These times can be used to provide an accurate clock in the MS and are also used to set the states of the various code generators in the MS.

Finally, the sync channel transmits information on the data rate used on the paging channel (4800 or 9600 bps).

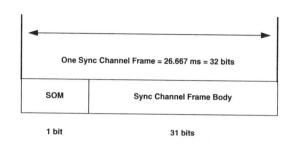

Figure 6.2 CDMA channel framing on forward sync channel.

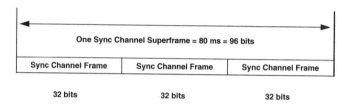

Figure 6.3 CDMA superframe structure on forward sync channel.

6.2.2 Paging Channel

The paging channel operates at a data rate of 4800 or 9600 bps and transmits overhead information, pages, and orders to an MS.

The Paging Channel message is similar in form to the Sync Channel message (fig. 6.1) and has an 8-bit message length header, a message body of a minimum of 2 bits and a maximum of 1146 bits, and a CRC code of 30 bits. The message length includes the header, body, and CRC, but not the padding. The CRC is computed on the message length header and the message body using the same code as the sync channel (equation 6.1).

Paging Channel messages can use synchronized capsules that end on a half-frame boundary or unsynchronized capsules that can end anywhere within a half-frame. If synchronized Paging Channel messages are less than an integer multiple of 47 bits for 4800-bps transmission (or 95 bits for 9600-bps transmission), they are padded with 0 bits at the end of the message. Unsynchronized messages do not have padding bits added to them.

After a message is formed, it is segmented into 47- or 95-bit chunks and sent in a sync channel half-frame (fig. 6.4) consisting of a 1-bit synchronized capsule indicator (SCI) field and 47 or 95 bits of the sync chan-

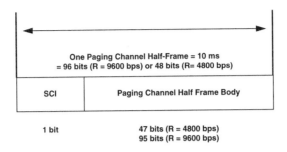

Figure 6.4 CDMA channel half-framing on forward paging channel.

nel frame body. A value of 1 for SCI indicates that the frame is the start of a Paging Channel message (either synchronized or unsynchronized). Messages can also start in the middle of a frame and immediately after the end of an unsynchronized message (with 0 padding bits). A value of 0 for SCI indicates that the frame is not the start of a message and can include a message (with or without padding), padding only, or the end of one message and the start of another.

Eight paging channel half-frames are combined to form a paging channel slot (fig. 6.5) of length 80 ms (384 bits at 4800 bps and 768 bits at 9600 bps). The entire Paging Channel message is then sent in N slots. The maximum number of slots that a message can use is 2048. The BS always starts a slot with a synchronized message capsule that starts at one bit after the beginning of a slot. The first bit in a slot is SCI = 1.

The paging channel sends many different types of messages; we mention a few and describe how they are used in chapter 7. We encourage you to consult the standards documents for a more detailed description of all of the messages. Some of the messages follow:

- **System Parameters Message:** This message is sent to all MSs in the area to specify the characteristics of the serving cellular/PCS system.
- **Access Parameters Message:** This message is sent to all MSs in the area to specify the characteristics of the messages sent on the access channel.
- **Order Message:** This message directs the MS to perform an operation and confirms a request from the MS. An example is an alerting message.

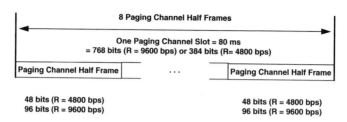

Figure 6.5 CDMA slot structure on the paging channel.

- **Channel Assignment Message:** This message informs the MS of the correct traffic channel to use for voice or data.
- **TMSI Assignment Message:** This message assigns a temporary mobile station identification (TMSI) to the MS. It is sent as part of the registration process described in chapter 7.

6.2.3 Traffic Channel

Channels not used for paging or sync can be used for traffic. The total number of traffic channels at a BS is 63 minus the number of paging and sync channels in operation at that BS.

Information on the traffic channels consists of primary traffic (voice or data), secondary traffic (data), and signaling in frames of length 20 ms. When the data rate on the traffic channel is 9600 bps, each frame of 192 bits consists of 172 information bits, 12 frame quality bits, and 8 encoder tail bits (set to all 0s). At 4800 bps, there are 80 information bits, 8 frame-quality bits, and 8 tail bits for a total of 96 bits. At 2400 and 1200 bps, there are 40 and 16 information bits and 8 tail bits, for a total of 48 and 24 bits, respectively. The BS can select the data transmission rate on a frame-by-frame basis. The data rate of 9600 bps can support multiplexed traffic and signaling. Data rates of 1200, 2400, and 4800 bps can support only primary traffic information. The receiving MS determines the data rate being received by a combination of symbol error rates at each data rate and the frame quality data at the higher data rates.

The frame quality indicator is a CRC on the information bits in the frame. At 9600 bps the generator polynomial is

$$g(x) = x^{12} + x^{11} + x^{10} + x^9 + x^8 + x^4 + x + 1 \qquad (6.2)$$

At 4800 bps, the generator polynomial is

$$g(x) = x^8 + x^7 + x^4 + x^3 + x + 1 \qquad (6.3)$$

At 9600 bps, the 172 information bits consist of 1 or 4 format bits and 171 or 168 traffic bits. A variety of different multiplexing options are supported. The entire 171 information bits can be used for primary traffic, or the 168 bits can be used for 80 primary traffic bits and 88 signaling traffic bits or 88 secondary traffic bits. Other options use 40 and 128 or 16 and 152 bits for primary and signaling/secondary traffic. Alternatively, the entire 168 bits can be used for signaling or secondary traffic.

When the forward traffic channel is used for signaling, the message is similar in form to the Paging Channel message (fig. 6.1) and has an 8-bit message length header, a message body of a minium of 16 bits and a maximum of 1160 bits, and a CRC code of 16 bits. Following the message are padding bits to make the message end on a frame boundary. The message length includes the header, body, and CRC, but not the padding. The CRC is computed on the message length header and the message body using the following code:

$$g(x) = x^{16} + x^{12} + x^5 + 1 \tag{6.4}$$

When the forward traffic channel is used for signaling, some typical messages that can be sent follow:

- **Order Message:** This is similar to the order message sent on the paging channel.
- **Authentication Challenge Message:** When the BS suspects the validity of the MS, it can challenge the MS to prove its identity. We examine this in more detail in chapter 7.
- **Send Burst Dual-Tone Multifrequency (DTMF):** When the BS needs dialed digits, it can request them in this message. This message would be used for digits for a three-way call, for example.
- **Extended Handoff Direction Message:** This message is one of several handoff messages sent by the BS. See chapter 7 for more details on the handoff process.

6.3 REVERSE CDMA CHANNEL

The MS communicates with the BS over the access channel or the reverse traffic channel. The access channel is used to make originations, process orders, and respond to pages. After voice or data communications are established, all communications occur on the reverse traffic channel.

6.3.1 Access Channel

Whenever an MS registers with the network, processes an order, sends a data burst, makes an origination, responds to a page, or responds to an authentication challenge, it uses the (reverse) access channel. The message on the reverse access channel consists of an access preamble of multiple frames of 96 zero bits with a length of 1 + PAM_SZ frames (fig. 6.6), followed by an access channel message capsule with length of 3 + MAX_CAP_SZ frames. The message capsule also consists of frames of length 96 bits. Since the data rate on the reverse access channel is 4800 bps, each frame has duration of 20 ms.

The entire access channel transmission therefore occurs in an access channel slot that has a length of

$$4 + \text{MAX\_CAP\_SZ} + \text{PAM\_SZ frames} \qquad (6.5)$$

where the values of MAX_CAP_SZ and PAM_SZ are received on the paging channel.

An access channel slot nominally begins at a frame where

$$t \bmod(4 + \text{MAX\_CP\_SZ} + \text{PAM\_SZ}) = 0 \qquad (6.6)$$

where t is the system time in frames.

The actual start of the transmission on the access channel is randomized to minimize collisions between multiple MSs accessing the channel at the same time.

All access channels corresponding to a paging channel have the same slot length. Different BSs may have different slot lengths.

The Access Channel message (fig. 6.7) is similar in form to the Sync Channel message and has an 8-bit message length header, a message body of a minimum of 2 bits and a maximum of 842 bits, and a CRC code of 30 bits. Following the message are padding bits to make the message end on a frame boundary. The message length includes the header, body, and CRC, but not the padding. The CRC is computed on the message length header and the message body using the same code as the sync channel (equation 6.1).

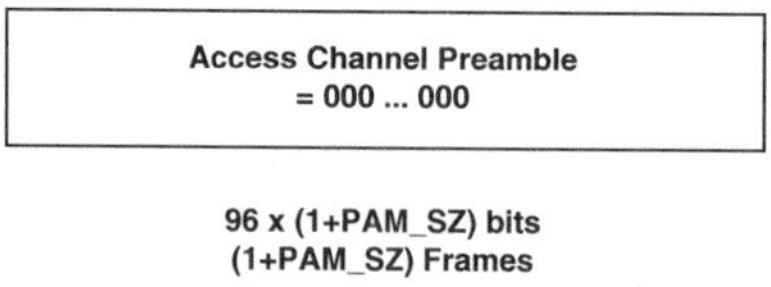

```
┌──────────────────────────────────┐
│       Access Channel Preamble     │
│            = 000 ... 000           │
│                                    │
└──────────────────────────────────┘

          96 x (1+PAM_SZ) bits
          (1+PAM_SZ) Frames
```

Figure 6.6 CDMA access channel preamble.

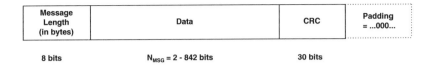

8 bits N_{MSG} = 2 - 842 bits   30 bits

Notes: N_{MSG} = Message length in bits (including length field and CRC)

Figure 6.7 CDMA message framing on access channel.

Each access channel frame contains either preamble bits (all zeros) or message bits. Frames containing message bits (fig. 6.8) have 88 message bits and 8 encoder tail bits (set to all zeros). Multiple frames are combined with an access channel preamble to form an access channel slot (fig. 6.9).

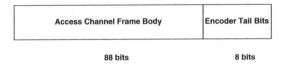

88 bits   8 bits

Figure 6.8 CDMA access channel framing.

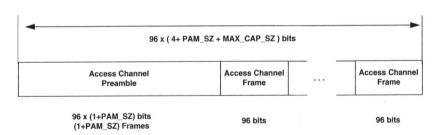

Figure 6.9 CDMA access channel slot.

6.3.2 Traffic Channel

Information on the reverse traffic channels consists of primary traffic (voice or data), secondary traffic (data), and signaling using frames of length 20 ms.

The message format is identical to the forward traffic channel. When the reverse traffic channel is used for signaling, the message (fig. 6.10) has an 8-bit message length header, a message body of a minimum

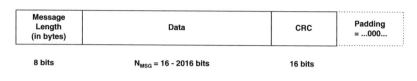

8 bits N_{MSG} = 16 - 2016 bits   16 bits

Notes: N_{MSG} = Message length in bits (including length field and CRC)

Figure 6.10 CDMA message framing on reverse traffic channel.

of 16 bits and a maximum of 2016 bits, and a CRC code of 16 bits. Padding bits follow the message to make the message end on a frame boundary. The message length includes the header, body, and CRC, but not the padding. The CRC is computed on the message length header and the message body using the code described in equation (6.4).

When the reverse traffic channel is used for signaling, some of the following example messages can be sent:

- **Order:** This message is either a response to a BS request or a request for service from the MS.
- **Authentication Challenge Response:** This message is sent in response to the challenge by the BS.
- **Flash with Information:** When the user requires special services from the BS, a flash message is sent. This messages is similar to depressing the switch-hook on a wireline phone. The message may or may not contain additional information.
- **Handoff Completion:** When the MS completes the handoff process, it sends this message.

6.4 FORWARD W-CDMA CHANNEL

The operation of the forward channel (BS to MS) in W-CDMA is similar to that of CDMA. However, the W-CDMA system separates the operation of layers 2 and 3 of the OSI protocol stack. Messages on the forward and reverse channel are formed at layer 3 and passed to layer 2. At layer 2, the messages are formed into a frame and sent to the physical layer (chapter 5).

The goal of the W-CDMA system is to model the operation of ISDN. Although the detailed ISDN message set is not used, the signaling data rates and voice-encoding rates are compatible with ISDN.

Data can be transmitted from a BS to an MS over the sync channel, paging channel, or information stream on the forward traffic channel.

Some of the data are specific to a particular channel. Other data (e.g., orders) can be sent on the paging channel or the traffic channel.

6.4.1 Layer-to-Layer Communications

The W-CDMA system defines a set of primitives for layer-to-layer communications to more closely implement the OSI reference model. Primitives are defined for layer 3-to-layer 2 communications and for layer 2-to-layer 1 communications. Primitives are also defined for management functions between layers. We examine the layer-to-layer communications here and refer you to the standard for the management functions.

In the CCITT layer control, four basic functions are defined for any primitive function (see fig. 6.11):

- **Request:** The higher layer makes a request to the lower layer. This request is passed to the receiving side.
- **Indication:** The lower layer at the receiving side passes an indication primitive to its next higher layer.
- **Response:** The higher layer on the receiving side performs an action (or requests an action from its higher layer) and sends a response message when the action is complete.
- **Confirm:** The lower layer passes the response to the transmitting side where the lower layer passes a confirmation (of the request) to its upper layer.

Not all primitives implement all four of the basic functions. Some implement only Request and Indication, for example.

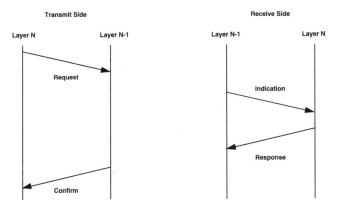

Figure 6.11 Interlayer primitives.

For layer 3-to-layer 2 communications, five primitives are defined (see table 6.1). Data link establish (DL-Establish) and release (DL-Release) are used to start and stop multiframe connection-oriented messages. Data link data (DL-Data) and Data link unit data (DL-Unit-Data) are used to transmit and receive acknowledged and unacknowledged messages, respectively. Data link setup (DL-Setup) is used to transmit and receive connectionless messages.

Table 6.1 Layer 3-to-Layer 2 Primitives

Primitive	Primitive Function			
	Request	Indication	Response	Confirm
DL-Establish	X	X	—	X
DL-Release	X	X	—	X
DL-Data	X	X	—	—
DL-Unit Data	X	X	—	—
DL-Setup	X	X	X	X

For layer 2-to-layer 1 communications, three primitives are defined (see table 6.2). Physical layer active (PH-Active) and physical layer deactivate (PH-Deactivate) are used to establish and release a physical layer connection. Physical data (PH-Data) is used to pass messages between layer 2 and layer 1.

Table 6.2 Layer 3-to-Layer 2 Primitives

Primitive	Primitive Function			
	Request	Indication	Response	Confirm
PH-Data	X	X	—	—
PH-Active	X	X	—	—
PH-Deactive	—	X	—	—

The layer 2-to-layer 3 primitives and the layer 2-to-layer 1 primitives are used in both the forward and reverse direction on the W-CDMA channel.

6.4.2 Sync Channel

The forward sync channel operates at a data rate of 16 kbps and transmits information that is specific to the BS and needed by the MS to access the system.

The Sync Channel message (fig. 6.12) has an 8-bit message length header, an 8- or 16-bit address field, an 8- or 16-bit control field, a layer 3 message body of a minimum of 0 bits and a maximum of 1232 bits, and a CRC code of 32 bits. If the Sync Channel messages are less than an integer multiple of 319 bits, they are padded with 0 bits at the end of the message. The message length includes the length field, the address field, the control field, the body, and CRC, but not the padding. The CRC is computed on the message length header, the address, the control, and the message body using the following code:

$$g(x) = x^{32} + x^{26} + x^{23} + x^{22} + x^{16} + x^{12} + x^{11} + x^{10} + x^8 + x^7 + x^5 + x^4 + x^2 + x + 1 \quad (6.7)$$

After a message is formed, it is segmented into 319-bits groups and sent in a sync channel frame (fig. 6.13) consisting of a 1-bit start of frame

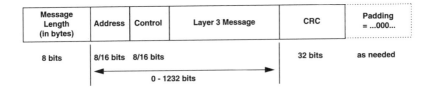

Notes: Message length (in bytes) includes: length field, address, control, layer 3 message and CRC)

Padding bits are not used for Unsynchronized Paging Channel Messages
Sync   Channel Data Rate = 16 kbps
Paging Channel Data Rate = 16 kbps

Figure 6.12 W-CDMA message framing on forward sync channel and paging channel.

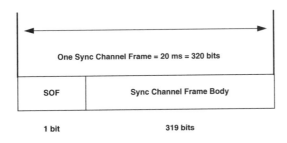

Note: SOF = 1 for first Body of Sync Channel Message,
        = 0 for all other Bodies in Sync Channel Message

Figure 6.13 W-CDMA channel framing on forward sync channel.

(SOF) field and 319 bits of the sync channel frame body. A value of 1 for SOF indicates that the frame is the start of a sync channel message. A value of 0 for SOF indicates that the frame is a continuation of a sync channel message or padding.

Four sync channel frames are combined to form a sync channel superframe (fig. 6.14) of length 80 ms (1280 bits). The entire sync channel message is then sent in N superframes. The padding bits are used so that the start message always starts at one bit after the beginning of a superframe. The first bit of the superframe is SOF = 1.

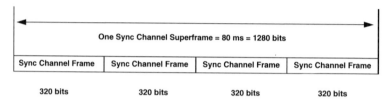

Figure 6.14 W-CDMA superframe structure on forward sync channel.

Two messages are sent on the sync channel: a mobility management (MM) identification message and a System Sync message. These two messages provide information similar to the single Sync Channel message of the CDMA system.

The System Sync message provides information about the number of 20-ms slots used for messages on the paging channel and the frequency allocation of the W-CDMA system (i.e., the bandwidth and channels used). The System Sync message defines the Walsh or Hadamard set used for the primary paging channel, the index into that set for the BS, and the pilot PN offset used by the BS. The message also informs the MS about the system date and time and the protocol revision supported by the BS.

The MM message provides the random number (RAND) used for authentication (see chapter 7), the system ID (SID), and the registration zone (REG_ZONE). The use of SID and REG_ZONE, for the purposes of paging roaming MSs, is discussed in chapters 4 and 7.

6.4.3 Paging Channel

For W-CDMA, the paging channel operates at a fixed data rate of 16 kbps and transmits overhead information, pages, and orders to a MS.

The Paging Channel message is identical in form to the sync channel message (fig. 6.12). The Paging Channel messages are sent in paging channel slots (fig. 6.15) of length 80 ms (1280 bits). The first bit of the slot is the start of slot (SOS) bit and the other 1279 bits contain the message. A group of 32 contiguous paging channel slots form a paging channel slot cycle of length 2.56 seconds (see fig. 6.16). The value of 1 for SOS indicates the start of a slot cycle. The other 31 slots in the cycle have value of 0 for SOS.

Paging Channel messages are sent in one or more slots. The message is padded to end on a slot boundary. Multiple short messages can be contained within a single slot or a long message can extend across multiple slots.

The types of messages sent on the forward paging channel are similar to those sent in the CDMA system. There are some exceptions; for example, the broadcast message is used to send information about the system and provides information similar to the system parameters message in the CDMA system. Chapter 7 describes the messages for the specific applications described there.

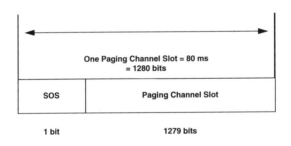

Figure 6.15 W-CDMA paging channel slots.

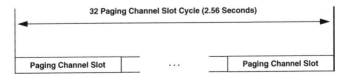

Figure 6.16 W-CDMA paging channel slot cycle.

6.4.4 Traffic Channel

The forward traffic channel operates at an effective data rate of 64 kbps. The signaling traffic, when sent, is at either 2 or 4 kbps. In an addition, there is a 2-kbps power control channel. The W-CDMA system supports voice or data traffic at rates of 16, 32, and 64 kbps. Chapter 7 discusses the voice coding for CDMA and W-CDMA. The entire traffic data stream of voice or data, signaling, and power control is multiplexed into a 64-kbps channel. The multiplexing is performed by a combination of the interleaver, symbol repetition, and multiplexer described in chapter 5 and figure 5.9.

When signaling is needed on the forward traffic channel, the messages that are sent are similar to those on the CDMA forward traffic channel. Specific examples of messages used are discussed in chapter 7.

6.5 REVERSE W-CDMA CHANNEL

The reverse W-CDMA functions are similar to those on the reverse CDMA channel but operate at a higher data rate. The MS communicates with the BS over the access channel or the reverse traffic channel. The access channel is used to make originations, process orders, and respond to pages. After voice or data communications are established, all communications occur on the reverse traffic channel.

The reverse channel implements the same layer 2 and layer 3 approach that is implemented on the forward channel.

6.5.1 Layer-to-Layer Communications

The same primitives described in section 6.4.1 for the forward channel are used for the reverse channel.

6.5.2 Access Channel

Whenever an MS registers with the network, processes an order, sends a data burst, makes an origination, responds to a page, or responds to an authentication challenge, it uses the (reverse) access channel.

The message on the reverse access channel consists of an access preamble of 1–3 frames of 320 zero bits (fig. 6.17), followed by an Access Channel message with length of 3 to 1 frames (fig. 6.18), for a total message length (fig. 6.19) of 4 frames (1280 bits). Since the data rate on the reverse access channel is 64 kbps, each frame has duration of 20 ms.

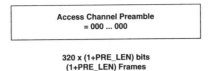

Figure 6.17 W-CDMA access channel preamble.

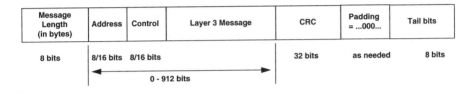

Notes: Message length (in bytes) includes: length field, address, control, layer 3 message and CRC)

Figure 6.18 W-CDMA message framing on access channel.

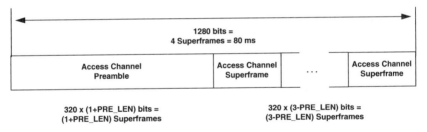

Figure 6.19 CDMA access channel slot.

The actual start of the transmission on the access channel is randomized to minimize collisions between multiple MSs accessing the channel at the same time.

The Access Channel message (fig. 6.18) is similar in form to the Sync Channel message and has an 8-bit message length header, an address field of 8 or 16 bits, a control field of 8 or 165 bits, a message body of a minimum of 0 bits and a maximum of 912 bits, and a CRC code of 32 bits. Following the message are padding bits and 8 encoder tail bits. The padding bits are added to make the message end on a frame boundary. The message length includes the header, address, control, body, and CRC, but not the padding and tail bits. The CRC is computed on the message length header and the message body using the same code as the sync channel equation (6.7).

Each access channel transmission consists of 1280 bits made up of four 320 bit frames (fig. 6.19). Each frame contains either preamble bits (all zeros) or message bits. Frames containing message bits (fig. 6.18) have 8 encoder tail bits (set to all 0s) at the end of the frame.

6.5.3 Traffic Channel

The format of the reverse traffic channel is identical to the forward traffic channel, except that the pilot signal is sent at reduced power.

The messages are similar to those on the reverse CDMA channel and are discussed in chapter 7.

6.6 SUMMARY

In this chapter, we examined the structure of the layer 2 and layer 3 messages on the forward (BS to MS) and reverse (MS to BS) channels for both the CDMA and W-CDMA systems. The CDMA system combines layers 2 and 3, whereas the W-CDMA system segments the functionality of layers 2 and 3. The forward channels are the sync channel, the paging channel, and the traffic channel. On the reverse link, the channels are the access channel and the traffic channel. The sync channel establishes the synchronization of the received data at the MS receiver. The paging channel is used to page the MS and send it orders. The reverse access channel enables the MS to respond to orders and originate calls. The traffic channel is used for digital voice or data. Both systems permit control information to be multiplexed into the traffic channel stream to send and receive data between the MS and the BS.

6.7 REFERENCES

1. TIA IS-95A, "Mobile Station–Base Station Compatibility Standard for Dual Mode Spread Spectrum Cellular System."
2. Alliance for Telecommunications Industry Standards J-STD-008, "Personal Station–Base Station Compatibility Requirements for 1.8 to 2.0 GHz Code Division Multiple Access (CDMA) Personal Communications Systems."
3. TIA IS-665, "W-CDMA (Wideband Code Division Multiple Access) Air Interface Compatibility Standard for 1.85-1.99 GHz PCS Applications."
4. Alliance for Telecommunications Industry Standards J-STD-015, "W-CDMA (Wideband Code Division Multiple Access) Air Interface Compatibility Standard for 1.85-1.99 GHz PCS Applications."

Signaling Applications in the CDMA System

7.1 INTRODUCTION

This chapter discusses the signaling applications in the CDMA system. The data application is discussed in chapter 10. As we described in chapter 5, the actual application is above the layer 7 software. All the U.S.-based cellular and PCS air interfaces share a common heritage from two sources. The original analog cellular system was defined in 1979 by the EIA. All the digital air interfaces, including CDMA and wideband CDMA, inherit their characteristics from the analog protocol [10]. In addition, they all have services based on the IS-41 standard for intersystem communications. IS-41 and IS-104 (for supplementary services) define the functionality for a mobile application part (MAP) for U.S.-based Pcss. They also use mobility management applications part (MMAP) for signaling from the base station to the mobile switching center. This signaling is based on ISDN.

In chapter 4, we discussed the TR-45/TR-46 model and the basic and supplementary services supported in a cellular/personal communication system. In this chapter, we will discuss the end-to-end call flow for some typical basic and supplementary services. These end-to-end call flows are synthesized from examination of the various standards and do not appear in any one document within the standards.

7.2 END-TO-END OPERATION OF A WIRELESS SYSTEM

This section describes the operation of a wireless system. We trace call flows from a mobile station to a base station to the MSC to other network elements. The flows are based on the TR-45/TR-46 reference model and an A-interface based on the Integrated Services Digital Network. The ISDN model assumes that there are ISDN terminals associated with the switch, one for each directory number on the switch. Since a cellular or personal communications system allows mobile stations to be associated with any base station, there is not a one-for-one correspondence between MSs and ISDN terminals. Thus, with the ISDN model, each MS registered at a base station is assigned a temporary directory number (also called virtual terminal number or interface directory number) that the radio system and PCS switch use to refer to the MS in the ISDN signaling messages, while that MS is registered at the base station. The ISDN A-interface defines a PCS application protocol (PCSAP) that uses ISDN signaling. With basic ISDN and PCSAP, the MSC can support terminal mobility. The radio system and switch can interact in either of two methods. In method one, the radio system is equivalent to a private branch exchange (PBX) and the switch is an ISDN switch that does minimal call control. In method two, the radio system is a virtual ISDN terminal with all call control in the ISDN switch. We have showed call flows for method two and leave the construction of call flows for method one as an exercise for you. Either method will work, although some systems designers will support one method over the other.

In chapter 3, we discussed the basic call-processing functions in the CDMA and wideband CDMA phones, and in chapter 6, we discussed some of the messages that are sent over the radio link. We will now use those functions and messages to describe the end-to-end call flows for several voice-based telecommunications services. In chapter 10, we will discuss wireless data services. We assume that you have studied the TR-45/TR-46 reference model described in chapter 4 and will refer to the various elements in that reference model as we describe the call flows.

7.2.1 Basic Services

Before a mobile station can originate or receive a call, it will register with the wireless system. An exception is made for emergency (911) calls. During the registration process, the MS is given a temporary mobile station identity (TMSI)[1] that is used for all subsequent call processing. We will first discuss the registration process and then call flows

---

1. Most cellular phones and some early PCS phones do not support TMSI and therefore will use a Mobile Identification Number (MIN) to identify themselves to the network.

for other services. Except for emergency calls, the mobile station must be registered on the PCS to receive services. The following call flows are based on an ISDN A-interface between the MSC and the BS. The call flows presented here are representative of the MS, BS, MSC, and PSTN network interactions. Many equipment vendors support proprietary interfaces between the MSC and BS. The call flows for their systems may be different within the network.

7.2.1.1 Registration   *Registration* is the means by which a mobile station informs a service provider of its presence in the system and its desire to receive service from that system. The MS may initiate registration for several different reasons.

A mobile station registering on an access channel may perform any of the following registration types:

- **Distance-Based Registration** is done when the distance between the current base station and the base station where the MS last registered exceeds a threshold.
- **Ordered Registration** is done when the system sets parameters on the forward paging channel to indicate that all or some of the MSs must register. The registration can be directed to a specific MS or a class of MSs.
- **Parameter Change Registration** is done when specific operating parameters in the MS are changed.
- **Power-Down Registration** is done when the mobile station is switched off. This allows the network to deregister a mobile station immediately upon its power-down.
- **Power-Up Registration** is done when power is applied to the mobile station and is used to notify the network that the MS is now active and ready to place or receive calls.
- **Timer-Based Registration** is done when a timer expires in the mobile station. This procedure allows the data base in the network to be cleared if a registered MS does not reregister after a fixed time interval. The time interval can be varied by setting parameters on the control channel.
- **Zone-Based Registration** is done whenever an MS enters a new area of the same system. A service area may be segmented into smaller regions, location areas, or zones, which are a group of one or more cells. The MS identifies the current location area via parameters transmitted on the forward paging channel. Location-based

registration reduces the paging load on a system by allowing the network to page only in the location area(s) where a mobile station is registered.

Two other forms of registration occur when the mobile station takes certain actions:

- **Implicit Registration** occurs when a mobile station successfully communicates with the base station for a page response or an origination.
- **Traffic Channel Registration** occurs when the mobile station is assigned a traffic channel. The base station can notify the MS that it is registered.

In chapter 3, we discussed the types of registration that CDMA and W-CDMA system implement.

The following steps are the call flows for the registration of all mobile stations listening to a paging channel (see fig. 7.1):

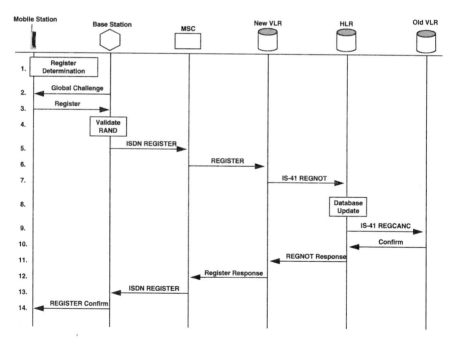

Figure 7.1 Mobile station registration.

1. The MS determines that it must register with the system.
2. The MS listens on the paging channel for the global challenge, RAND.
3. The MS sends a message to the base station with its international mobile station identification (IMSI), RAND, response to the challenge (AUTHR), and other parameters, as needed, in the MS registration request.
4. The base station validates RAND.
5. The base station sends an ISDN REGISTER message to the MSC.
6. The MSC receives the REGISTER message and sends a message to the serving VLR.
7. If the MS is not currently registered to the serving VLR, the VLR sends a REGistration NOTification (REGNOT) message to the user's HLR containing the IMSI, and other data as needed.
8. The MS's HLR receives the REGNOT message and updates its data base accordingly (stores the location of the VLR that sent the REGNOT message).
9. The MS's HLR sends an IS-41 REGistration CANCel (REG-CANC) message to the old VLR where the MS was previously registered so that the old VLR can cancel the MS's previous registration.
10. The old VLR returns a confirmation message that includes the current value of call history count (CHCNT).
11. The user's HLR then returns a REGNOT response message to the (new) visited VLR and passes along information that the VLR needs (e.g., user's profile, interexchange carrier ID, shared secret key for authentication, and current value of CHCNT). If the registration is a failure (due to invalid IMSI, service not permitted, nonpayment of bill, etc.), then the REGNOT response message will include a failure indication.
12. Upon receiving the REGNOT response message from the user's HLR, the VLR assigns a temporary mobile station identification (TMSI) and then sends a registration notification response message to the MSC.
13. The MSC receives the message, retrieves the data, and sends an ISDN REGISTER message to the base station.
14. The base station receives the REGISTER message and forwards it to the MS to confirm registration.

Some CDMA systems will support the sending of the old TMSI when an MS registers in a new system. When an MS sends its old TMSI, the call process flow is similar except that the new VLR communicates with the old VLR to obtain the IMSI before an HLR query can be done.

7.2.1.2 Call Origination *Call origination* is the service wherein the MS user calls another telephone on the world-wide telephone network. It is a cooperative effort among the MSC, the VLR, and the base station.

The detailed call flow steps follow (see fig. 7.2 for the call flow diagram):

1. The MS processes an Origination Request from the user and sends it to the base station.
2. The base station sends a PCSAP Qualification Request to the VLR.
3. The VLR returns a Qualification Request Response to the base station.

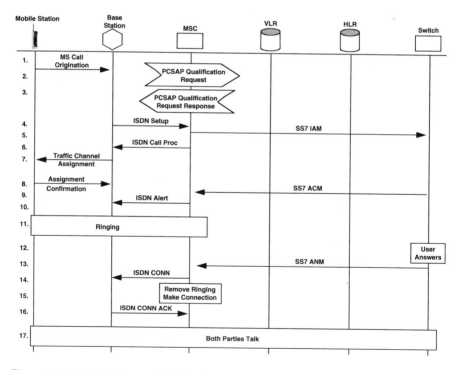

Figure 7.2 Mobile station call origination.

4. The base station then processes an ISDN Setup message and sends it to the MSC.

5. The MSC sends an SS7 (ISDN User Part) Initial Address message (IAM) to the terminating switch (wireline or wireless).

6. At the same time, the MSC returns an ISDN Call Proceeding message to the base station.

7. The base station assigns a traffic channel to the MS.

8. The MS tunes to the traffic channel and confirms the traffic channel assignment.

9. The terminating switch checks the status of the called telephone and returns an SS7 Address Complete message (ACM) to the MSC.

10. The MSC returns an ISDN alert message to the base station.

11. The MSC provides audible ringing to the user.

12. The terminating user answers.

13. The terminating switch sends an SS7 ANswer message (ANM) to the MSC.

14. The MSC sends and ISDN CONNect message to the base station.

15. The MSC removes audible ringing and makes the network connection.

16. The base station returns an ISDN CONNect ACKnowlege message.

17. The two parties establish their communications.

7.2.1.3 Call Termination   *Call termination* is the service wherein an MS user receives a call from other telephones in the world-wide telephone network. The following discussion is for calls terminating to a MS registered at its home MSC. Calls terminating to roaming MSs will be discussed in section 7.2.1.5. Call termination is a cooperative effort among the MSC, the VLR, and the base station.

The detailed call flow steps follow (see fig. 7.3 for the call flow diagram):

1. A user in the world-wide phone network (wired or wireless) dials the directory number (DN) of the MS.

2. The originating switch sends an SS7 IAM to the MSC.

3. The MSC queries the VLR for the list of radio systems (one or more) where the MS will be paged and for the TMSI of the MS.

4. The VLR returns with the TMSI and a list of base stations.

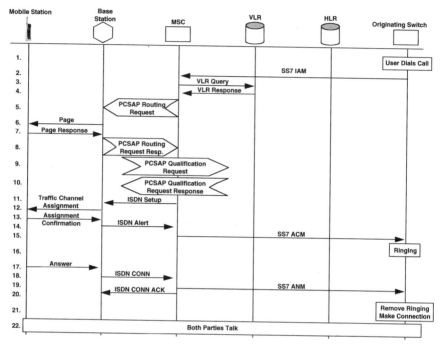

Figure 7.3 Call termination to a mobile station.

5. The MSC sends an PCSAP Routing Request message to all base stations on the list.

6. Each base station broadcasts a Page message on appropriate paging channels.

7. The MS responds to the page with a Page Response message at one base station.

8. The base station sends a PCSAP Routing Request Response to the MSC.

9. The base station sends a PCSAP Qualification Directive message to the VLR.

10. The VLR responds with a PCSAP Qualification Request Response.

11. The MSC sends an ISDN Setup message to the base station.

12. The base station sends a traffic channel assignment to the MS.

13. The MS tunes to the traffic channel and sends a Traffic Channel Assignment Confirmation message.

14. The base station sends an ISDN Alert message to the MSC.

15. The MSC sends an SS7 ACM to the originating switch.
16. The originating switch applies audible ringing to the network.
17. The user answers, and the MS sends a response message to the base station.
18. The base station sends an ISDN CONNect message to the MSC.
19. The MSC sends an SS7 ANswer Message (ANM) to the originating switch.
20. The MSC sends and ISDN CONNect ACKnowlege message to the base station.
21. Audible ringing is removed.
22. The two parties establish their communications.

7.2.1.4 Call Clearing When either party in a conversation wishes to end a call, then the call clearing function is invoked. The exact call flows depend on which side ends the call first. It is a cooperative effort among the MSC, the VLR, and the base station.

The detailed call flow steps, for an MS-initiated call clearing follow (see fig. 7.4 for the call flow diagram):

1. The MS user hangs up.

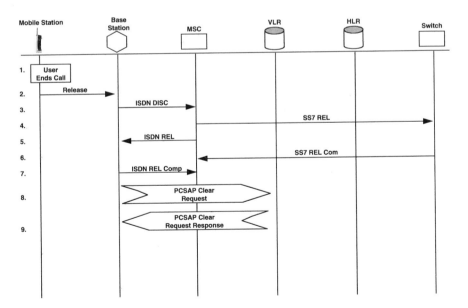

Figure 7.4 Call clearing—mobile station initiated.

2. The MS sends a Release message to the base station.

3. The base station sends an ISDN DISConnect message to the MSC.

4. The MSC sends an SS7 RELease message to the other switch.

5. The MSC sends an ISDN RELease message to the base station.

6. The other switch sends an SS7 RELease Complete message to the MSC.

7. The base station sends an ISDN RELease Complete message to the MSC.

8. The base station sends a PCSAP Clear Request message to the VLR.

9. The VLR closes the call records and sends a PCSAP Clear Request Response message to the base station.

The detailed call flow steps, for a far end–initiated call clearing follow (see fig. 7.5 for the call flow diagram):

1. The far end user hangs up.

2. The far end switch sends an SS7 RELease message to the MSC.

3. The MSC sends an ISDN DISConnect message to the base station.

4. The base station sends a Release message to the MS.

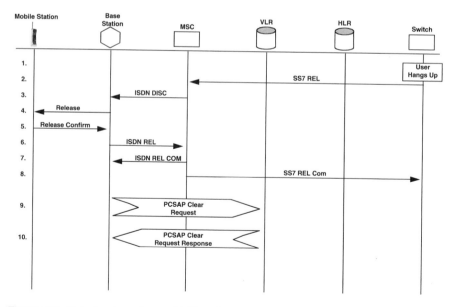

Figure 7.5 Call clearing—far end initiated.

5. The MS confirms the message and disconnects from the traffic channel.
6. The base station sends an ISDN RELease message to the MSC.
7. The MSC sends an ISDN RELease COMplete message to the base station.
8. The MSC sends an SS7 RELease COMplete message to the other switch.
9. The base station sends a PCSAP Clear Request message to the VLR.
10. The VLR closes the call records and sends a PCSAP Clear Request Response message to the base station.

7.2.1.5 Roaming   *Roaming* is the ability to deliver services to mobile stations outside of their home area. When an MS is roaming, registration, call origination and call delivery will take extra steps. Whenever data will be retrieved from the VLR, and the data are not available, then the VLR will send a message to the appropriate HLR to retrieve the data. The data consist of IMSI-to-MIN conversion, service profiles, shared secret data (SSD) for authentication, and other data needed to process calls. The most logical time to retrieve this data is when the MS registers with the system.

Once the data on a roaming MS are stored in the VLR, then call processing for any originating services (basic or supplementary) is identical to that of home MSs. However, there may be times when the MS originates a call before registration has been accomplished or when the VLR data are not available. At those times, an extra step will be added for the VLR to retrieve the data from the HLR. Thus any originating service has two optional steps where the VLR sends a message (using IS-41 signaling over SS7) to the HLR requesting data on the roaming MS. The HLR will return a message with the proper call information.

Call delivery is not possible to an unregistered MS because the network does not know where the MS is located. When the MS is registered with a system, call delivery to the roaming MS is possible. This section will discuss call delivery to roaming MSs in detail.

There are two cases of call delivery to roaming MS:

1. The MS has a geographic-based directory number (indistinguishable from a wireline number) and
2. The MS has a nongeographic number.

The call flows for both operations will be described.

When the MS has a geographic number, the MSC is assigned a block of numbers that are within the local numbering plan for the area of the world where the MSC is located. Call routing to the MS is then done according to the procedures for that of a wireline telephone.[2] If an MS associated with a MSC is not in its home area, the MSC will query the HLR for the location of the MS. The MSC then invokes call forwarding to the MSC where the MS is located, and the connection is made to the second MSC where call-terminating services are delivered according to the procedures in section 7.2.1.3. This procedure is inefficient because it results in two sets of network connections: originating switch to home MSC and home MSC to visited MSC.

Call delivery to a roaming MS is a cooperative effort among the home and visited MSC, the VLR and HLR, and the radio system. The detailed call flow steps for call delivery to a roaming MS with a geographic directory number follow (see fig. 7.6 for the call flow diagram):

1. A user in the world-wide phone network (wired or wireless) dials the directory number of the MS.

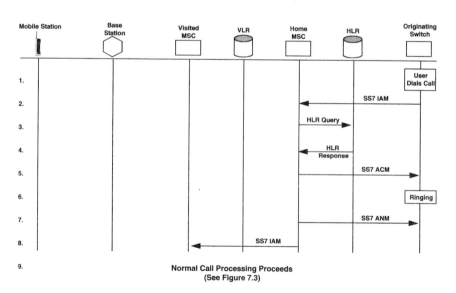

Figure 7.6 Call termination to a roaming mobile station with a geographic number.

2. For example, in New Jersey, 908-313-XXXX is used by the local cellular provider for cellular phones in Elizabeth, NJ. The wireline network routes calls to those numbers in a normal fashion and calls terminate on the cellular switch.

2. The originating switch sends an SS7 IAM to the home MSC.

3. The home MSC queries the HLR for the location of the MS.

4. The HLR returns the location of the visited system.

5. The home MSC sends an SS7 ACM message to the originating switch.

6. Ring is applied.

7. The home MSC sends an SS7 answer message to the originating switch.

8. The MSC invokes call forwarding to the MSC in the visited system and the forwarding (home) MSC switch sends an SS7 IAM to the visited MSC.

9. Call processing proceeds at step 3 of the terminating call flow.

When the MS has a nongeographic number, calls can be directed from an originating switch directly to the visited switch. Call delivery to a nongeographic number requires that the originating switch recognize the number as a nongeographic number and do special call processing for routing. This special processing is known as intelligent network (IN) processing. If the originating switch does not support IN, then it will route the call to a switch that supports IN. With IN support, the originating switch will recognize the nongeographic number and send an SS7 message to the HLR with a request for the location of the MS. The HLR will return a temporary directory number (on the visited MSC) that can be used to route to the MS in the visited system. Calls then proceed according to normal terminating call flows.

Call delivery to a roaming MS, with a nongeographic number is, therefore, a cooperative effort among the visited MSC, the VLR and HLR, and the radio system. The detailed call flow steps for call delivery to a roaming MS with a nongeographic directory number follow (see fig. 7.7 for the call flow diagram):

1. A user in the world-wide phone network (wired or wireless) dials the directory number of the MS.

2. The originating switch recognizes the number as a nongeographic number and sends an SS7 Query message to the HLR at the home MSC.

3. The HLR returns the location of the visited system with a directory number to use for further call processing.

4. The originating switch sends an SS7 IAM to the visited MSC.

5. Call processing proceeds at step 3 of the terminating call flow.

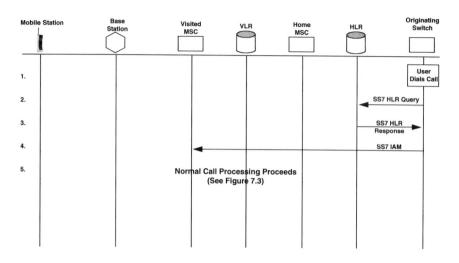

Figure 7.7 Call termination to a roaming mobile station with a nongeographic number.

7.2.2 Unique Challenge

There are several fraud problems with wireless phones [8, 9, 10] and the standards support security features of global challenge, SSD update, and unique challenge. We discuss the unique challenge here, and you can consult the standards for a description of other security features.

The unique challenge protects the network from fraudulent use by illegal mobile stations. At various times throughout a call, the network may want to challenge the validity of a mobile station communicating with the network. If the radio link communications are encrypted, it is unlikely that someone may have stolen the radio link from a legitimate user. The stealing of the radio link is called hijacking the link. Only those systems that operate unencrypted or have encryption disabled because of system overloads, national emergencies, or other reasons are subject to hijacking from illegal mobile stations.

The unique challenge can be sent to a MS at any time. It is typically initiated by the MSC in response to some event (registration failure and after a successful handoff are the most typical cases). The following steps are the call flow for a unique challenge (see fig. 7.8):

1. The MSC decides to perform a unique challenge.
2. The MSC sends a PCSAP message to the base station with TMSI (or MIN or IMSI if the MS is not registered) and RANDU.

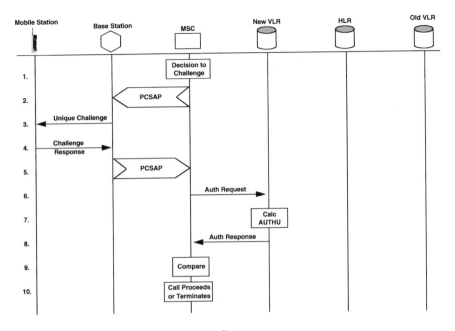

Figure 7.8 Shared secret key unique challenge.

3. The base station forwards the unique challenge message to the MS.

4. The mobile station calculates its specific response to the unique challenge (AUTHU) and sends to the base station a Unique Challenge Order Response message that includes TMSI (or MIN or IMSI), AUTHU, and other data as needed.

5. The base station forwards the message to the MSC in the PCSAP message.

6. The MSC sends an Authentication Request message to the VLR with TMSI (or MIN or IMSI), RAND, and AUTHU and requests that the VLR perform the same calculation as done by the MS.

7. The VLR checks its data base for TMSI (or MIN or IMSI). If the data are not in the VLR, the VLR queries the HLR for the data. When the data are in the data base at the VLR, it calculates the value of AUTHU.

8. The VLR returns a message to the MSC.

9. The MSC compares the AUTHU from the MS and from the VLR.

10. The MSC decides to continue or interrupt call processing. If the two AUTHUs match, then the MSC continues call processing. If

they do not match, then the MSC optionally may take action (e.g., terminate a call in progress or deregister the MS).

7.2.3 Supplementary Services

IS-41 supports several supplementary services (see section 4.3.2); however, only the call flow for call waiting is described herein. For other services, see either the standards or *Wireless and Personal Communications Systems* [9].

Call waiting provides notification to a wireless subscriber of an incoming call while the user's mobile station is in the busy state. Subsequently, the user can either answer or ignore the incoming call. Once the call is answered, the user can switch between the calls until one or more parties hang up. When either distant party hangs up, then the call reverts to a normal (non-call-waiting call). If the MS user hangs up, then both calls are cleared according to normal call-clearing functions.

The detailed call flow steps for call-waiting delivery to a mobile station follow (see fig. 7.9 for the call flow diagram):

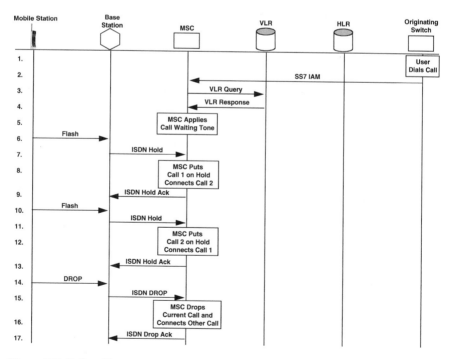

Figure 7.9 Call waiting.

1. User dials a call.
2. The originating switch sends an SS7 IAM to the MSC.
3. The MSC queries the VLR.
4. The VLR returns with a location of the MS that is within the serving system. If it is not, then the call is forwarded to the serving MSC.
5. The MSC determines that the MS is busy and subscribes to call waiting and thus applies a call-waiting tone.
6. The user presses the flash button (it may be SEND on some MSs) to answer the call-waiting indication, and the MS sends a Flash message to the base station.
7. The base station sends and ISDN Hold message to the MSC.
8. The MSC puts the first call on hold and connects the second call.
9. The MSC sends an ISDN Hold Acknowledge to the base station.
10. The User presses the flash button (it may be SEND on some MSs) to talk to caller 1, and the MS sends a Flash message to the base station.
11. The base station sends an ISDN Hold message to the MSC.
12. The MSC puts the second call on hold and connects the first call.
13. The MSC sends an ISDN Hold Acknowledge to the base station.
14. The user wants to drop the current call (either 1 or 2) and pushes the drop (or END) key and the MS sends a DROP message to the base station.
15. The base station sends an ISDN Drop message to the MSC.
16. The MSC drops the current call and connects the other call (the one currently on hold).
17. The MSC sends an ISDN Drop Acknowledge message to the base station.

7.2.4 Handoffs

A wireless telephone (mobile station) moves around a geographic area. When the station is idle, it periodically reregisters with the system according to the parameters described in section 7.2.1.1. When a call is active, the combination of the mobile station, the base station, and the MSC manage the communications between the base station and mobile station so that good radio link performance is maintained. The process whereby a mobile station moves to a new traffic channel is called *handoff*. The original analog cellular system processed handoffs by commanding the mobile station to tune to a new frequency. For analog cellular, the handoff process caused a short break in the voice path and a noticeable "click" was heard by both parties in the telephone call. For data modems, the click often causes data errors or loss of data synchronization.

For the CDMA (and wideband CDMA) systems, the characteristics of the spread spectrum communications permit the system to receive the mobile transmissions on two or more base stations simultaneously. In addition, the mobile station can simultaneously receive the transmissions of two or more base stations. With these capabilities, it is possible to process a handoff from one base station to another, or from one antenna face to another on the same base station, without any perceptible disturbance in the voice or data communications.

During handoff, the signaling and voice information from multiple base stations must be combined (or bridged) in a common point with decisions made on the "quality" of the data. Similarly, voice and signaling information must be sent to multiple base stations, and the mobile station must combine the results. The common point could be anywhere in the network but is typically at the MSC. The call flows described here for handoff assume that the MSC contains the bridging circuitry.

The CDMA system defines several types of handoffs.

- **Soft Handoff** occurs when the new base station begins communications with the mobile station while the mobile station is still communicating with the old base station. The network (MSC) combines the received signals from both base stations to process an uninterrupted signal to the distant party. The mobile station will receive the transmissions from the two base stations as additional multipaths in the RAKE receiver and will process them as one signal.

- **Softer Handoff** occurs when the mobile station is in handoff between two different sectors at the same base station. Typically, a base station is designed so that an antenna transmits and receives over a 60° or 120° sector rather than a full 360°. For full (360°) coverage, multiple base station antennas are then needed. For the purposes of discussion of softer handoffs, it is useful to designate a sector as a primary sector (i.e., the oldest sector serving the call). Since an MS will typically communicate only with three base stations during a soft handoff, only one (or none) softer handoff can be associated with a call at any particular time.

- **Hard Handoff** occurs when the two base stations are not synchronized or are not on the same frequency and an interruption in voice or data communications occurs. Hard handoffs can occur when more than one frequency band is used, or the two base stations are not synchronized (e.g., there are in two different systems).

Another type of hard handoff occurs when there is no serving CDMA base station available and the mobile station must be directed to an analog cellular channel. In this book, we are discussing digital transmissions; however, during the transition time from analog to digital, there will be mixed systems in existence, and some mobile stations may be capable of both digital and analog operation. For more details, consult chapter 4, and the standards [4, 5, 6].

- **Semisoft Handoff** occurs when the handoff appears as a soft handoff within the network but the mobile station processes it as a hard handoff.

In CDMA, both the base station and the mobile station monitor the performance of the radio link and can request handoffs. Handoffs requested by a mobile station are called *mobile-assisted handoffs,* and those requested by the base station are called *base station-assisted handoffs.* Either side can initiate the handoff process whenever the following triggers occur:

- **Base Station Traffic Load.** The network can monitor loads at all base stations and trigger handoffs to balance loads between them to achieve higher traffic efficiency.
- **Distance Limits Exceeded.** Since all base stations and mobile stations are synchronized, both sides can determine base to mobile range. When the distance limit is exceeded, either side can request a handoff.
- **Pilot Signal Strength Below Threshold.** When the received signal strength of the pilot signal falls below a threshold, either side can initiate a handoff.
- **Power Level Exceeded.** When the base station commands a mobile station to increase its power and the maximum power level of the mobile station is exceeded, then either side can request a handoff.

The mobile station determines the parameters for the handoff request from the system parameters message in the CDMA system and the broadcast message in the wideband CDMA system. Both messages are transmitted on their systems' paging channels.

As we have described, the handoff process is a cooperative effort among the old and new base stations, the mobile station, and the MSC. The following call flows are based on a frame relay A-interface [12] between the base station and the MSC. The call flows are included as representative calls flows. The actual calls flows may be either standard or proprietary to an equipment vendor.

The detailed call flow steps for a CDMA soft handoff (beginning) follow (see fig. 7.10 for the call flow diagram):

1. The mobile station determines that another base station has a sufficient pilot signal to be a target for handoff.
2. The mobile station sends a Pilot Strength Measurement message to the serving base station.
3. The serving base station sends an Interbase Station Handoff Request message to the MSC.
4. The MSC accepts the Handoff Request and sends an Interbase Station Handoff Request message to the target base station.

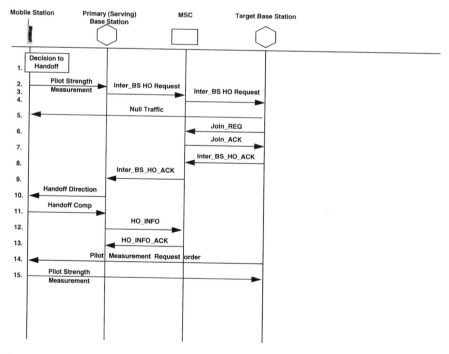

Figure 7.10 CDMA soft handoff—beginning.

5. The target base station establishes communication with the mobile station by sending it a Null Traffic message.

6. The target base station sends a Join Request message to the MSC.

7. The MSC conferences the connections from the two base stations so that the handoff can be processed without a break in the connection (i.e., soft handoff) and sends a Join Acknowlege message to the target base station.

8. The target base station sends an Interbase Station Handoff Acknowledgment message to the MSC.

9. The MSC sends a Interbase Station Handoff Acknowledgment message to the serving base station.

10. The serving base station sends a Handoff Direction message to the mobile station.

11. The mobile station sends a Handoff Complete message to the serving base station.

12. The new serving station sends a Handoff Information message to the MSC.

13. The MSC confirms the message with a Handoff Information Acknowledgment message.

14. The target base station sends a Pilot Measurement Request Order to the mobile station.

15. The mobile station sends a Pilot Strength Measurement message to the target base station.

The mobile unit is now communicating with two base stations (i.e., it is in soft handoff). Both base stations must communicate with the MSC. MSC uses the highest quality signals from the two base stations and sends transmitted signals to both base stations.

After the mobile station is in soft handoff, one of the signals may fall below a predetermined threshold (based on information sent in overhead messages on the control channel), and the mobile stations will request that one base station be removed from the connection. The detailed call flow steps for a CDMA soft handoff with the serving base station dropping off follow (see fig. 7.11 for the call flow diagram):

1. The mobile station determines that the serving base station has insufficient pilot signal to continue to be a base station in the soft handoff.

2. The mobile station sends a Pilot Strength message to the serving base station. The message requests that the base station drop off from the handoff.

3. The serving base station sends a Handoff Direction message to the mobile station. The message indicates which base station to be dropped from the soft handoff (in this case, the serving base station).

4. The mobile station sends a Handoff Complete message to the serving base station.

5. The serving base station sends an Interface Primary Transfer message to the target base station with relevant call record information.

6. The target base station confirms the message with an Interface Primary Transfer Acknowledge message.

7. The target base station then sends a Handoff Information message to the MSC.

8. The MSC sends a Handoff Information Acknowlege message to the target base station.

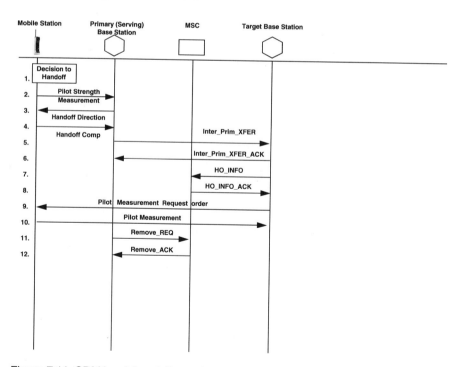

Figure 7.11 CDMA soft handoff: serving base station dropping off.

9. The target base station sends a Pilot Measurement Request Order to the mobile station.

10. The mobile station sends a Pilot Strength Measurement message to the target base station.

11. The old serving base station sends to the MSC a Remove Request message, which requests that the base station be dropped from the connection.

12. The MSC confirms the message by sending a Remove Acknowledge message to the old base station.

The mobile station is now communicating with the target base station (new serving base station). If additional soft handoffs are needed, the handoff beginning procedure is repeated.

The procedures to drop a target base station from a soft handoff are similar to those that drop the serving base station. The detailed call flow steps for a CDMA soft handoff with the target base station dropping off follow (see fig. 7.12 for the call flow diagram):

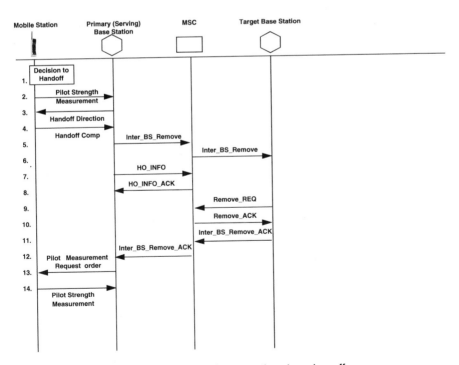

Figure 7.12 CDMA soft handoff: target base station dropping off.

1. The mobile station determines that the target base station has insufficient pilot signal to continue to be a base station in the soft handoff.
2. The mobile station sends a Pilot Strength message to the serving base station. The message requests that the target base station drop off from the handoff.
3. The serving base station sends a Handoff Direction message to the mobile station, which indicates the base station is to be dropped from the soft handoff (in this case, the target base station).
4. The mobile station sends a Handoff Complete message to the serving base station.
5. The serving base station sends an Interbase Station Remove message to the MSC.
6. The MSC sends an Interbase Station Remove message to the appropriate base station (in this case, the target base station).
7. The serving base station then sends a Handoff Information message to the MSC.
8. The MSC sends a Handoff Information Acknowlege message to the serving base station.
9. The target base station sends a Remove Request message to the MSC.
10. The MSC sends a Remove Acknowledge message to the target base station.
11. After the target base station removes its resource from the call, it sends an Interbase station Remove Acknowledge message to the MSC.
12. The MSC sends a Remove Acknowledge message to the serving base station.
13. The serving base station sends a Pilot Measurement Request Order to the mobile station.
14. The mobile station sends a Pilot Strength Measurement message to the serving base station.

The mobile station is now communicating only with the serving base station. If additional soft handoffs are needed, the handoff beginning procedure is repeated.

7.3 SUMMARY

In this chapter, we have discussed the signaling applications for a CDMA wireless telephony system. Since end-to-end call flows are not presented in any of the standards but are distributed across several standards, we described several basic and supplementary call flows. First we examined the registration process. Because an unregistered phone cannot place or receive calls, it is necessary for a mobile station to register on the network. We described the call flow for registration including the authentication procedures used to validate the identity of the mobile station. After a phone is registered, it can place or receive calls; therefore, we described the call flows for a mobile-originated call and a mobile-terminated call. When a call is in progress, either the mobile station or the far end can end the call and release the connection. We examined the call flows for both release procedures.

An important component of wireless services is the ability to find and place calls to a roaming mobile station. Even though most mobile stations have a geographic number now, many will have nongeographic numbers in the future, so we described call flows for both. Geographic numbers are phone numbers that are located to a specific point on the world-wide phone system. Nongeographic numbers do not have a location associated with them, and the network must maintain a data base of the location of the phone. Additional routing steps are necessary to place a call to a mobile station with a nongeographic number.

Because fraud is a problem in the analog AMPS system, CDMA and W-CDMA implement cryptographic methods for combating fraud. At any time during call processing, the MSC can present a unique challenge to a mobile station to confirm its identity. We describe the call flow without revealing details that would permit the procedure to be defeated.

Even though cellular and personal communications systems (and CDMA, in particular) have a rich set of supplementary features, we examine the most common feature of call waiting. The various standards describe additional procedures for all the basic and supplementary services described in chapter 4. We encourage you to consult the standards [1–7, 11, 12] for additional information.

Finally, since the CDMA system processes handoffs differently than analog cellular or TDMA cellular/PCS systems, we describe the soft handoff process for CDMA and present call flows for soft handoff beginning and ending.

7.4 REFERENCES

1. Committee T1—Telecommunications, A Technical Report on Network Capabilities, Architectures, and Interfaces for Personal Communications, T1 Technical Report #34, May 1994.

2. Committee T1, Stage 2 Service Description for Circuit Mode Switched Bearer Services, Draft T1.704.

3. TIA Interim Standard IS-41 C, "Cellular Radiotelecommunications Intersystem Operations."

4. TIA IS-95A, "Mobile Station–Base Station Compatibility Standard for Dual Mode Spread Spectrum Cellular System."

5. Alliance for Telecommunications Industry Standards J-STD-008, "Personal Station–Base Station Compatibility Requirements for 1.8 to 2.0 GHz Code Division Multiple Access (CDMA) Personal Communications Systems."

6. TIA IS-665, "W-CDMA (Wideband Code Division Multiple Access) Air Interface Compatibility Standard for 1.85–1.99 GHz PCS Applications."

7. Alliance for Telecommunications Industry Standards J-STD-015, "W-CDMA (Wideband Code Division Multiple Access) Air Interface Compatibility Standard for 1.85–1.99 GHz PCS Applications."

8. Report of the Joint Experts Meeting on Privacy and Authentication for PCS, November 8–12, 1993, Phoenix, Arizona.

9. Garg, V. K., and Wilkes, J. E., *Wireless and Personal Communications Systems*, Prentice Hall, Upper Saddle River, NJ, 1996.

10. Wilkes, J. E., "Privacy and Authentication Needs of PCS," *IEEE Personal Communications*, 2 (4), August 1995, pp. 11–15.

11. EIA/TIA-553, "Cellular System Mobile Station—Land Station Compatibility Specification."

12. TR-45 Contribution, "Frame Relay A-Interface."

Voice Applications in the CDMA System

8.1 INTRODUCTION

This chapter discusses the voice-coding application in the CDMA system. The data application is discussed in chapter 10. Since voice encoding is critical to digital transmission and is the primary application that most wireless phone users need, we discuss the voice-coding algorithms used by both CDMA and wideband CDMA. CDMA uses an 8-kbps (or a 13.2-kbps) data rate for voice transmission, whereas wideband CDMA uses either ADPCM at 32 kbps or PCM at 64 kbps to provide a service more closely modeled on the current wireline network.

8.2 VOICE ENCODING

The wireline network is based on sending voice using digital pulse code modulation (PCM) at 64 kbps and sending data at rates of 64 kbps or higher multiples of 64 kbps. Many older analog facilities still exist, especially in residential areas, and use voice band modems at rates up to 28.8 kbps for data and analog electrical signals for voice. At the central office, the analog voice and analog data are converted to digital signals using PCM or, optionally, using modem pools for data.

It would be optimal if the identical systems could be used for wireless communications. Unfortunately the error rates on the radio channels are many orders of magnitude higher than those of the copper or fiber optic cables. In addition, PCM is inefficient for use over scarce and expensive radio channels.

Therefore, both the CDMA and wideband CDMA systems use some efficient method of voice coding and extensive error recovery techniques to overcome the harsh nature of the radio channel. The CDMA system uses a code-excited linear prediction (CELP) voice-encoding system at 8 kbps and optionally at 13 kbps; the wideband CDMA system uses ADPCM at 32 kbps as its primary voice-coding system and PCM at 64 kbps as an option.

In this section we will cover the various means used for both of these protocols to send voice signals over the radio channel.

8.2.1 Pulse Code Modulation

The simplest form of waveform coding scheme is linear PCM, in which the speech signal is band-limited, compressed, sampled, quantized, and encoded (see fig. 8.1). This approach is widely used for analog-to-digital conversion of a signal. In radio and telephone communications, it is not necessary to send the entire 20–20,000 Hz signal normally used for high-fidelity music. Intelligible speech communications can occur with a much narrower bandwidth and, therefore, more efficient range of frequencies. For telephone communications, the speech signal is band-limited to a frequency range 300–3300 Hz. To achieve telephone quality speech, 12 bits per sample are required at a sampling rate of 8000 samples per second. However, by using a logarithmic sampling system, 8 bits per sample are sufficient. Each sample is then quantized into one of 256 levels. Telephone speech uses two widely different variations of PCM to achieve quality speech (μ-law and A-law PCM). Both are based on a non-uniform quantization of the signal amplitude according to a logarithmic scale rather than a linear scale.

The decoder for PCM (fig. 8.2) inverts the stages of the encoding process. PCM encoding and decoding are inherently simple systems. However, they require a high bit rate for transmission.

For PCM, North America and Japan use μ-law encoding where the output digital signal $s(t)$ is related to the input signal $i(t)$ by

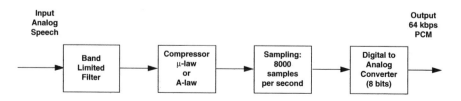

Figure 8.1 PCM encoder.

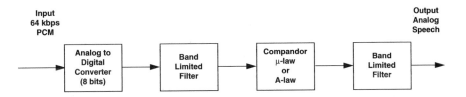

Figure 8.2 PCM decoder.

$$s(t) = \sin(i(t))\frac{\ln(1 + \mu|i(t)|)}{\ln(1 + \mu)}, \quad -1 \le i(t) \le 1 \qquad (8.1)$$

where a typical value for $\mu = 255$ is used in the United States.

In equation (8.1), the input signal is normalized to a range of ±1. It can be noted that for small $i(t)$, $s(t)$ approaches a linear function and that for large $i(t)$, $s(t)$ approaches a logarithmic function. The purpose of the μ-law encoding is to improve the signal-to-noise ratio for weak speech signals. The overall data rate is 64 kbps with sampling at 8 kbps and 8 bits per sample.

In Europe, PCM uses A-law encoding where the output digital signal $s(t)$ is related to the input signal $i(t)$ by

$$s(t) = \sin(i(t))\frac{1 + \ln(A|i(t)|)}{1 + \ln A} \qquad -\frac{1}{A} \le |i(t)| \le 1 \qquad (8.2)$$

$$s(t) = \sin(i(t))\frac{1 + (A|i(t)|)}{1 + \ln A} \qquad 0 \le |i(t)| \le \frac{1}{A}$$

Where a typical value of $A = 87.6$ is used in Europe.

In equation (8.2), the input signal is also normalized to a range of ±1. Note that $s(t)$ is logarithmic for $|i(t)| < 1/A$ and linear for $|i(t)| > 1/A$. Thus, A-law provides a somewhat flatter signal-to-distortion performance compared to μ-law when the signal is greater than $1/A$, at the expense of poorer performance at low signal levels.

Telephone communications that cross continental borders must have conversion routines in their transmission paths if the two continents use different encoding laws.

8.2.2 Adaptive Differential Pulse Code Modulation

High bit rates are not attractive for wireless systems since the capacity of the system is low. Higher system capacities are obtained with differential coders, where compression can be applied dynamically, such as adaptive predictive coding (APC) and adaptive differential pulse code modulation (ADPCM). The reason for these coders is to achieve a better signal-to-quantization noise performance and a lower coding rate over PCM.

Differential coders generate error signals, as the difference between the input speech samples and corresponding prediction estimates. The error signals are quantized and transmitted. ADPCM and APC differential coders are often used for intermediate bit rate between 16 and 32 kbps.

ADPCM employs a short-term predictor that models the speech spectral envelope. It achieves network-quality speech (mean opinion score [MOS] of 4.1 or better) at 32 kbps. This is a low-complexity, low-delay coder of reasonable robustness with channel bit error rates in the range of 10^{-3} to 10^{-2}. The ADPCM coder is well suited for wireless access applications.

In an ADPCM encoder (fig. 8.3), first the analog speech is converted to PCM. If the signal is already PCM, from the network for example, then the analog to the PCM step is not needed. The A-law or μ-law encoded signal is then converted to a uniform PCM level (i.e., equal steps between levels) signal. The encoder generates a difference signal between the converted signal and an estimated signal and encodes the estimated signal using 15 levels. The resultant signal is transmitted at 32 kbps (half the rate for PCM). In the encoder, the signal estimator is generated by an inverse quantizer and an adaptive predictor. The use of differential signals and proper design of the predictor enables an overall coding efficiency improvement over PCM.

In the ADPCM decoder (fig. 8.4), the input 32-kbps signal is processed by a inverse adaptive quantizer and an adaptive predictor. The

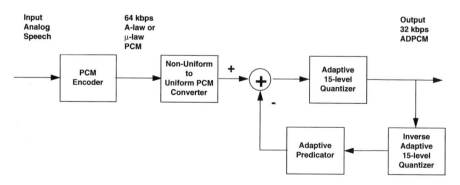

Figure 8.3 ADPCM encoder. (Reproduced under written permission of the copyright holder [TIA].)

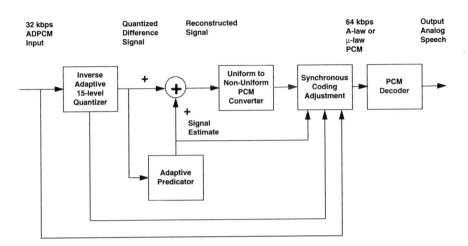

Figure 8.4 ADPCM decoder. (Reproduced under written permission of the copyright holder [TIA].)

output of the quantizer and the predictor are combined to generate a reconstructed signal that is converted back to PCM. The regenerated PCM signal is then processed (in the synchronous coding adjustment stage) with signals from the input, the quantizer output, and the predictor output to generate the A-law or μ-law PCM signal. The processing in the synchronous coding adjustment stage ensures that the PCM signal is correctly modeled by converting it to uniform PCM and comparing the error signals that result from the actual received signal. If an error occurs, it is corrected before the output PCM signal is generated. The ADPCM signal conforms to CCITT recommendation G.721. Finally, if analog speech is needed, the signal is processed by a PCM decoder.

The IS-665 standard supports a modified version of ADPCM that has been optimized for the PCS environment. The functional operation is the same as ADPCM; the coding algorithm has been modified. Refer to the IS-665 standard for the details.

8.2.3 Code-Excited Linear Predictor

PCM, ADPCM, and APC operate in the time domain. No attempt is made to understand or analyze the information that is being sent. To achieve lower coding rates, redundancy removal techniques operating in the frequency domain have been used successfully. Frequency domain waveform coding algorithms decompose the input speech signal into sinusoidal components with varying amplitudes and frequencies. Thus, the speech is modeled as a time-varying line spectrum. Frequency domain coders are systems of moderate complexity and operate well at a medium bit rate (16 kbps). When designed to operate in the range of 4.8–9.6 kbps, the complexity of the approach used to model the speech spectrum increases considerably.

The other class of speech-coding techniques consists of algorithms called *vocoders*, which attempt to describe the speech production mechanism in terms of a few independent parameters serving as the information-bearing signals. These parameters attempt to model the creation of the voice by the vocal tract, decompose the information, and send it to the receiver. The receiver attempts to model an electronic vocal tract to produce the speech output.

The modeling operates in the following way. Vocoders consider that speech is produced from a source-filter arrangement. Voiced speech is the result of exciting the filter with a periodic pulse train (simulating the opening and closing of the vocal cords). Unvoiced speech is the result of exciting the filter with random noise (simulating air rushing past a constriction in the vocal tract). Vocoders operate on the input signal using an analysis process based on a particular speech production model and extract a set of source-filter parameters that are encoded and transmitted. At the receiver, they are decoded and used to control a speech synthesizer, which corresponds to the model used in the analysis process. Provided that all the perceptually significant parameters are extracted, the synthesized signal, as perceived by human ear, resembles the original speech signal. Nonspeech signals are often not modeled well, so this method works poorly for analog modems.

Vocoders are medium-complexity systems and operate at low bit rates, typically 2.4 kbps, with synthetic-quality speech. Their poor-quality speech is due to the oversimplified source model used to drive the filter and the assumption that the source and filter are linearly independent.

In the bit rates, from about 5 to 16 kbps, the best speech quality is obtained by using hybrid coders, which use suitable combinations of waveform-coding and vocoder techniques. A simple hybrid coding scheme for telephone-quality speech with a few integrated digital signal processors is the residual excited linear prediction (RELP) coding. This belongs to a class of coders known as an analysis-synthesis coder based on linear predictive coding (LPC).

The RELP systems employ short-term (and in certain cases, long-term) linear prediction to formulate a difference signal (residual) in a feed-forward manner. RELP systems are capable of producing communications quality speech at 8 kbps. These systems use either pitch-aligned, high-frequency regeneration procedures or full-band pitch prediction in time domain to remove the pitch information from the residual signal prior to band-limitation/decimation. At bit rates less than 9.6 kbps, the quality of the recovered speech signal can be improved significantly by an analysis by synthesis (AbS) optimization procedure to define the excitation signal. In these systems, both the filter and the excitation are defined on a short-term basis using a closed-loop optimization process that minimizes a perceptually weighted error measure formed between the input and decoded speech signals.

CDMA uses a variation of RELP called code-excited linear prediction. With this technique, the CELP decoder (fig. 8.5) uses a codebook to generate inputs to a synthesis filter. The codebook is characterized by its codebook index I and gain G. The spectral filter is characterized by three sets of parameters: the pitch spectral lines a, the pitch lag L, and the pitch gain b. The output of the filter is processed by a post filter and gain adjustment.

CDMA implements a rate 1 encoder at 8.55 kbps and supports rates of 4, 2, and 0.8 kbps (rates 1/2, 1/4 and 1/8, respectively). Each of the rates uses successively less bits for encoding the values of I, G, L, b, and a. At rate of 1/8 (fig. 8.6), insufficient bits are available to send the codebook index I, and a pseudorandom code generator (synchronized at both ends) is used and seeded by a random seed of value CBSEED.

The basic frame for CDMA is 20 ms. At rate 1, 160 bits are sent for encoding the data plus an 11-bit parity check field. Fewer bits are used at lower data rates (see table 8.1).

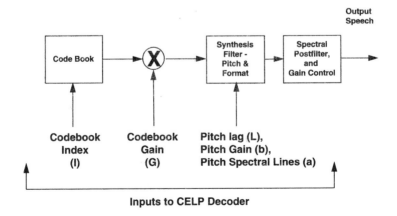

Figure 8.5 CELP decoder for rates 1, 1/2, and 1/4.

Table 8.1 CELP Parameters for Various Coding Rates

CELP Parameters	Rate 1	Rate 1/2	Rate 1/4	Rate 1/8
Line spectral pairs i bits	40	20	10	10
i updates per frame	1	1	1	1
Total i bits per frame	40	20	10	10
Pitch lag L bits	7	7	7	0
L updates per frame	4	2	1	0
Total L bits per frame	28	14	7	0
Pitch gain b bits	3	3	3	0
b updates per frame	4	2	1	0
Total b bits per frame	12	6	3	0
Codebook index I bits	7	7	7	0
I updates per frame	8	4	2	—
Total I bits per frame	56	28	14	0
Codebook gain G bits	3	3	3	2
G updates per frame	8	4	2	1
Total G bits per frame	24	12	6	2
Codebook seed CBSEED bits	0	0	0	4
CBSEED updates per frame	—	—	—	1
Total CBSEED bits	—	—	—	4
Parity check bits per frame	11	0	0	0
Total number of bits per frame	171	80	40	16

The CELP speech encoder requires three steps to implement. First the line spectral pairs (LSP) i values are determined. Then the LSP values are used in an analysis by synthesis process to determine the values for the pitch lag L and gain b. Finally, the values of i, L, and b are used in a second AbS step to determine the codebook indices I and gains G. We now describe these steps in more detail.

- **LSP Determination** (fig. 8.7). The encoder for the LSP codes first converts the speech to uniform PCM with at least 14 bits. If the

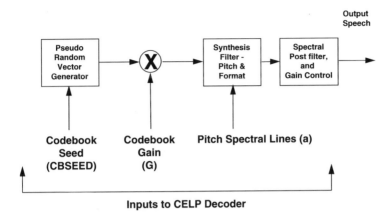

Figure 8.6 CELP decoder for rate of 1/8.

encoder is in a base station, then the received speech is most likely µ-law PCM; if the encoder is in a mobile station, then the received speech is analog. After the speech is converted to PCM, it is processed to remove the DC component and filtered by a Hamming window. The autocorrelation of the sampled output is then com-

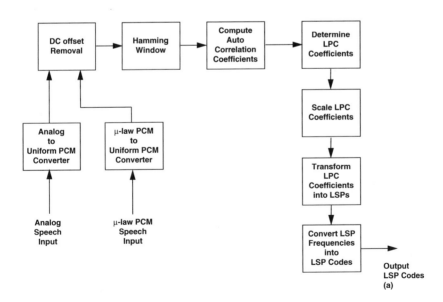

Figure 8.7 CELP encoder for LSP codes.

puted and used to determine the coefficients for the linear predictive coding. The LPC coefficients are then scaled, transformed into the frequency components, and converted into the values for the i bits of the coder output.

• **The Pitch Lag and Gain Bits** (fig. 8.8). They are computed by a recursive process where the output of the PCM encoder is combined with the LSP codes previously calculated and with all possible values of pitch and gain. For each value of pitch and gain, an error function is computed, and the transmitted values for pitch and gain are chosen to minimize the error.

• **The Codebook Index and Gain** (fig. 8.9). They are computed in a similar recursive process using the uniform PCM signal; the computed values for frequency, pitch, and gain; and all possible codebook values and gains.

For the rate of 1/8, codebook indices are not computed, but a random vector generated at both sides is used.

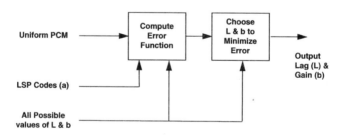

Figure 8.8 CELP encoder for pitch parameters.

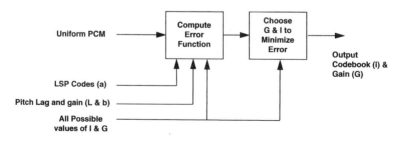

Figure 8.9 CELP encoder for codebook values.

8.3 SUMMARY

This chapter discussed the digital voice-encoding systems used for both CDMA and wideband CDMA. Conventional wireline systems transmit voice by digitizing the voice signal using PCM at a rate of 64 kbps. While it is possible to use PCM in wireless systems (the W-CDMA system uses it as an option), the capacity of the wireless system is lower compared to using other digitizing methods for voice. The two CDMA systems use different approaches to digitizing the voice signal. The CDMA system uses a CELP at 8 or 13.2 kbps to digitize voice. CELP systems model the operation of the human vocal tract to code speech efficiently. We described the operation of the CELP encoder and decoder at a high level. The wideband CDMA system uses a version of PCM called adaptive differential PCM at 32 kbps. To understand ADPCM, we first described PCM and then the ADPCM system. The W-CDMA system uses a modified version of ADPCM. Throughout the descriptions on voice coding, our goal has been to explain the coding systems at a high level so that you can understand the operation of the system and read the standards with some understanding of the motivation for them. Those of you who need to design systems or want additional information are encouraged to read the standards.

8.4 REFERENCES

1. TIA IS-96A, "Speech Service Option Standard for Wideband Spread Spectrum Digital Cellular System."
2. TIA IS-665, "W-CDMA (Wideband Code Division Multiple Access) Air Interface Compatibility Standard for 1.85–1.99 GHz PCS Applications."
3. Recommendation G162, CCITT Plenary Assembly, Geneva, May–June 1964, Blue Book, Vol. 111, P. 52.
4. ITU Recommendation G.711.

RF Engineering and
Facilities Engineering

9.1 INTRODUCTION

This chapter presents basic guidelines for engineering a CDMA system. The topic is extremely complex and cannot be covered extensively in a single chapter. Moreover, many of the effective techniques are currently being developed as CDMA is deployed in the commercial market. This chapter discusses several topics that are germaine to the engineering of a CDMA system: propagation models, link budgets, the transition from analog operation to CDMA operation, facilities engineering, radio link capacity, and border cells on an boundary between two service providers.

9.2 RADIO DESIGN FOR A CELLULAR/PCS NETWORK

Many factors need to be considered early in the design of a cellular/PCS network for a metropolitan urban area. For example, the extent of radio coverage for indoor locations, the quality of service for different environments, efficient use of the spectrum, and the evolution of the network are some of the key factors that need to be carefully evaluated by all prospective service providers. Often, these factors are further complicated by the constraints imposed by the operating environments and regulatory issues. A system designer must carefully balance all the trade-offs to ensure that the network is robust, future-proof, and of high service quality.

9.2.1 Radio Link Design

For any wireless communications system, the first important step is to design the radio link. This is required to determine the base station density in different environments as well as the corresponding radio coverage. For a wireless network to provide good quality indoor and outdoor service in an urban environment, flexibility and resilience should be incorporated into the design. The transmit power of handsets will be the determining factor for a CDMA system with balanced uplink/downlink power.

Although the mobile antenna gain does not affect the balancing of the link budget, it is an important factor in the design of the power budget for handset coverage. From a user's point of view, a cellular/PCS network should imply that there is little restriction on making or receiving calls within a building or traveling in a vehicle using handsets. A system should be designed to allow the antenna of a handset to be placed in non-optimal positions. In addition, the antenna may not even be extended when calls are being made or received. In normal system designs, it is assumed that the gain of a mobile antenna is 0 dBi.[1] However, allowing for the handset antennas to be placed in suboptimal positions, a more conservative gain of –3 dBi should be used. In reality, the antenna gain, because of the positioning of an antenna in an arbitrary position or with the antenna retracted into the handset housing, could be as low as –6 to –8 dBi, depending on specific handsets and their corresponding housing designs.

9.2.2 Coverage Planning

The most important design objective of a cellular/PCS network is to provide a near-ubiquitous radio coverage. One of the most important considerations in the radio coverage planning process is the propagation model. The accuracy of the prediction by a particular model depends on its ability to account for the detailed terrain, vegetation, and buildings. This accuracy is of vital importance to determine the path loss and, hence, the cell sizes and the infrastructure requirement of a cellular/PCS network. An overestimation will lead to an inefficient use of the network resources, whereas an underestimation will result in poor radio coverage. Propagation models generally tend to oversimplify real-life propagation conditions and may be grossly inaccurate in complex metropolitan

---

1. dBi refers to the gain relative to an isotropic antenna.

urban environments. The empirical propagation models provide general guidelines only, and they are too simplistic for accurate network design. Accurate field measurements must be made to provide information on the radio coverage in an urban environment. Measured data can be used either directly in the planning process to access the feasibility of individual cell site or indirectly to calibrate the coefficients of the empirical propagation model to achieve better characterization of a specific environment.

Radio propagation in an urban environment is subject to shadowing. To ensure that the signal level in 90 percent of the cell area is equal to or above the specified threshold, a shadow-fading margin, which is dependent on the standard deviation of the signal level, must be included in the link budget. For a typical urban environment, a shadow-fading margin of 8–9 dB should be used based on the assumption that the path loss follows an inverse 2-5 exponent law, i.e., the path loss is inversely proportional to the distance of separation raised to a power between 2 and 5. The value of the power is dependent upon propagation characteristics.

Another critical factor that affects the radio coverage is the penetration loss for both buildings and vehicles. If the radio coverage for the outer portion of a building is sufficient, then an assumed penetration loss of 10–15 dB should be adequate. However, if calls are expected to be received and originated within the inner core of the building, a penetration loss of about 30 dB should be used. Similarly, for in-vehicle coverage, the penetration loss is equally important. A car could experience a penetration loss of 3–6 dB, whereas vans and buses have even larger variations. The penetration loss at the front of a van should be no more than that experienced in a car, but the loss at the back of a van could be as high as 10–12 dB, depending on the amount of window space. Thus, for design purposes, a high penetration loss should be assumed to ensure a good service quality. For an urban environment, as building penetration loss is the dominant factor, in-vehicle penetration will generally be sufficient as a consequence.

9.3 PROPAGATION MODELS

Propagation models are used to determine how many base stations are required to provide the coverage requirements needed for the network. Initial network design typically engineers for coverage. Later growth of network design engineers for capacity. Some systems may need to start with wide area coverage and high capacity and, therefore, may start at a later stage of growth.

The coverage requirement is coupled with the traffic-loading requirements, which rely on the propagation model chosen to determine the traffic distribution and the off-loading from an existing base station to new base stations as part of a capacity relief program. The propagation model helps to determine where the base stations should be located to achieve an optimal position in the network. If the propagation model used is not effective in helping to place base stations correctly, the probability of incorrectly deploying a base station into the network is high.

The performance of the network is affected by the propagation model chosen because it is used for interference predictions. As an example, if the propagation model is inaccurate by 6 dB, then E_b/N_0 could be 13 or 1 dB (assuming that $E_b/N_0 = 7$ dB is the design requirement). Based on traffic-loading conditions, designing for a high E_b/N_0 level could negatively affect financial feasibility. On the other hand, designing for a low E_b/N_0 would degrade the quality of service.

The propagation model is also used in other system performance aspects including handoff optimization, power level adjustments, and antenna placements. Although no propagation model can account for all perturbations experienced in the real world, using one or more models for determining the path losses in the network is essential. Each of the propagation models being used in the industry has pros and cons. It is through a better understanding of the limitations of each of the models that a good RF engineering design can be achieved in a network.

9.3.1 Modeling for the Outside Environment

9.3.1.1 Analytical Model The propagation loss between the base station and the mobile station in the outside environment has been extensively studied. The propagation loss is generally expressed by the following expression [3,4]:

$$P(R) = N(R, \sigma) + n \log \frac{R}{R_0} \qquad (9.1)$$

where $P(R)$ = loss at distance R relative to the loss at a reference distance R_0,

 n = path loss exponent,

 σ = standard deviation, typically 8 dB.

The second term on the right-hand side of equation (9.1) represents a constant attenuation in the outside environment between the base station and the mobile station. Typically, n approximately equals 4, although it may range between 2 (which equals the loss in free space) and 5. If n is equal to 4, then the signal will be attenuated 40 dB if the distance increases 10 times with respect to the reference distance. The first term in equation (9.1) represents the variation in the loss about the average path loss. This function is an approximate log-normal distribution with an average equal to the second term and a standard deviation of approximately 8 dB. It has been found that this value is applicable for a wide range of radio environments, including urban and rural areas.

9.3.1.2 Empirical Models Several empirical models have been suggested and used to predict propagation path losses. We discuss the two widely used models—the Hata-Okumura model and the Walfisch-Ikegami model.

The Hata-Okumura Model [1]. Most of the propagation tools use a variation of Hata's model. Hata's model is an empirical relation derived from the technical report made by Okumura [6] so that the results could be used in computational tools. Okumura's report consists of a series of charts that have been used in radio communication modeling. The following are the expressions used in Hata's model in order to determine the mean loss L_{50}:

Urban area:

$$L_{50} = 69.55 + 26.16 \log f_c - 13.82 \log h_b - a(h_m) +$$
$$(44.9 - 6.55 \log h_b) \log R \text{ dB} \tag{9.2}$$

where f_c = frequency (MHz),
$\quad\quad L_{50}$ = mean path loss (dB),
$\quad\quad h_b$ = base station antenna height (m),
$\quad\quad a(h_m)$ = correction factor for mobile antenna height (dB),
$\quad\quad R$ = distance from base station (km).

The range of the parameters for which Hata's model is valid is

$$150 \le f_c \le 1,500 \text{ MHz,}$$
$$30 \le h_b \le 200 \text{ m,}$$
$$1 \le h_m \le 10 \text{ m,}$$
$$1 \le R \le 20 \text{ km.}$$

$a(h_m)$ is computed as:

For a small or medium-sized city:

$$a(h_m) = (1.1 \log f_c - 0.7)h_m - (1.56 \log f_c - 0.8) \text{ dB} \qquad (9.3)$$

For a large city:

$$a(h_m) = 8.29(\log 1.54 \ h_m)^2 - 1.1 \text{ dB}, \qquad f_c \leq 200 \text{ MHz} \qquad (9.4)$$

or

$$a(h_m) = 3.2(\log 11.75 \ h_m)^2 - 4.97 \text{ dB}, \qquad f_c \geq 400 \text{ MHz} \qquad (9.5)$$

Suburban Area:

$$L_{50} = L_{50}(urban) - 2\left[\left(\log\left(\frac{f_c}{28}\right)^2\right) - 5.4\right] \text{ dB} \qquad (9.6)$$

Open Area:

$$L_{50} = L_{50}(urban) - 4.78(\log f_c)^2 + 18.33 \ (\log f_c) - 40.94 \text{ dB} \qquad (9.7)$$

Hata's model does not account for any of the path-specific correction used in Okumura's model.

Okumura's model [6] tends to average over some of the extreme situations and does not respond sufficiently quickly to rapid changes in the radio path profile. The distance-dependent behavior of Okumura's model is in agreement with the measured values. Okumura's measurements are valid only for the building types found in Tokyo. Experience with comparable measurements in the United States has shown that a "typical" United States suburban situation is often somewhere between Okumura's suburban and open areas. Okumura's suburban definition is more representative of residential metropolitan area with large groups of "row" houses.

Okumura's model requires that considerable engineering judgment be used, particularly in the selection of the appropriate environmental factors. Data are needed in order to be able to predict the environmental factors from the physical properties of the buildings surrounding a mobile receiver. In addition to the appropriate environmental factors, path-specific corrections are required to convert Okumura's mean path loss predictions to the predictions that apply to the specific path under study. Okumura's techniques for correction of irregular terrain and other path-specific features require engineering interpretations and are thus not readily adaptable for computer use.

The Walfisch/Ikegami Model [12].   This model[2] is used to estimate the path loss in an urban environment for cellular communication. The model is a combination of the empirical and deterministic model for estimating the path loss in an urban environment over the frequency range of 800–2000 MHz. This model is used primarily in Europe for the GSM system and in some propagation models in the United States. The model contains three elements: free space loss, roof-to-street diffraction and scatter loss (see fig. 9.1), and multiscreen loss. The expressions used in this model are

$$L_{50} = L_f + L_{rts} + L_{ms} \qquad (9.8)$$

or

$$L_{50} = L_f \text{ when } L_{rts} + L_{ms} \le 0 \qquad (9.9)$$

where L_f = free space loss,
L_{rts} = roof-to-street diffraction and scatter loss,
L_{ms} = multiscreen loss.

Free space loss is given as:

$$L_f = 32.4 + 20 \log R + 20 \log f_c \text{ dB} \qquad (9.10)$$

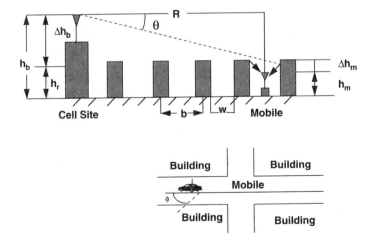

Figure 9.1 The Walfisch-Ikegami propagation model.

---

2. This model is also known as the Cost 231 model.

The roof-to-street diffraction and scatter loss is given as

$$L_{rts} = (-16.9) - 10 \log W + 10 \log f_c + 20 \log \Delta h_m + L_0 \text{ dB} \quad (9.11)$$

where W = street width (m),
$\Delta h_m = h_r - h_m$ (m).
L_0 = -9.646 dB $\quad 0 \le \phi \le 35$ degree
L_0 = $2.5 + 0.075(\phi - 35)$ dB $\quad 35 \le \phi \le 55$ degree
L_0 = $4 - 0.114(\phi - 55)$ dB $\quad 55 \le \phi \le 90$ degree
where ϕ = incident angle relative to the street

The multiscreen loss is given as

$$L_{ms} = L_{bsh} + k_a + k_d \log R + k_f \log f_c - 9 \log b \quad (9.12)$$

where b = distance between buildings along the radio path (m)
L_{bsh} = $-18 \log 11 + \Delta h_b$ $\quad h_b > h_r$,
L_{bsh} = 0 $\quad h_b < h_r$,
k_a = 54 $\quad h_b > h_r$,
k_a = $54 - 0.8 h_b$ $\quad R \ge 500$ m, $h_b \le h_r$,
k_a = $54 - 1.6 \Delta h_b R$ $\quad R < 500$ m, $h_b \le h_r$.

Note: Both L_{bsh} and k_a increase the path loss with lower base station antenna heights.

$$k_d = 18 \quad h_b < h_r,$$

$$k_d = 18 - \frac{15 \Delta h_b}{\Delta h_m} \quad h_b \ge h_r,$$

$$k_f = 4 + 0.7 \left(\frac{f_c}{925} - 1 \right) \quad \text{for midsize city and suburban area with moderate tree density,}$$

$$k_f = 4 + 1.5 \left(\frac{f_c}{925} - 1 \right) \quad \text{for metropolitan center.}$$

The range of parameters for which Walfisch-Ikegami model is valid follows:

$$800 \le f_c \le 2,000 \text{ MHz}$$

$$4 \le h_b \le 50 \text{ m}$$

$$1 \le h_m \le 3 \text{ m}$$

$$0.02 \le R \le 5 \text{ km}$$

The following default values can be used for the model:

b = 20 – 50 m,
W = $b/2$,
ϕ = 90 degree,
Roof = 3 m for pitched roof and 0 m for flat roof,
h_r = 3 (number of floors) + Roof.

Using the following data, a comparison of the path loss from the Hata and Walfisch-Ikegami models is made in table 9.1.

f_c = 880 MHz,
h_m = 1.5 m,
h_b = 30 m,
Roof = 0 m,
h_r = 30 m,
ϕ = 90 degree,
b = 30 m,
W = 15 m.

Table 9.1 A Comparison of Path Loss from Hata and Walfisch-Ikegami Model

Distance, km	Path Loss, dB	
	Hata's Model	Walfisch–Ikegami Model
1	126.16	139.45
2	136.77	150.89
3	142.97	157.58
4	147.37	162.33
5	150.79	166.01

The path losses predicted by Hata's model are 13–16 dB lower than those predicted by the Walfisch–Ikegami model. Hata's model ignores effects from street width, street diffraction, and scatter losses, which the Walfisch–Ikegami model includes.

Correction Factor for Attentuation Due to Trees. Weissberger [13] has developed a modified exponential delay model that can be used where a radio path is blocked by dense, dry, in-leaf trees found in temperate climates. The additional path loss can be calculated from the following expression:

$$L_f = 1.33(f_c)^{0.284}(d_f)^{0.588} \text{ dB} \quad \text{for } 14 \le d_f \text{ m} \quad (9.13)$$

$$= 0.45(f_c)^{0.284}d_f \text{ dB} \quad \quad \text{for } 0 \le d_f \le 14 \text{ m}$$

where L_f = loss in dB,
$\quad\quad f_c$ = frequency in GHz,
$\quad\quad d_f$ = tree height in meters.

The difference in path loss for trees with and without leaves has been found to be about 3–5 dB. For a frequency of 900 MHz, equation (9.13) is reduced to

$$L_f = 1.291(d_f)^{0.588} \text{ dB} \quad\quad \text{for } 14 \leq d_f \leq 400 \text{ m} \quad\quad (9.14)$$

$$= 0.437 d_f \text{ dB} \quad\quad \text{for } 0 \leq d_f \leq 14 \text{ m}$$

9.3.2 Models for Indoor Environment

Experimental studies have indicated that a portable receiver moving in a building experiences Rayleigh fading for obstructed propagation paths and Ricean fading for line of sight (LOS) paths, regardless of the type of building. *Rayleigh fading* is short-term fading resulting from signals traveling separate paths (multipath) that partially cancel each other. A LOS path is clear of building obstructions; in other words, there are no reflections of the signal. *Ricean fading* results from the combination of a strong LOS path and a ground path plus numerous weak reflected paths. A TIA IS-95A mobile station, however, is capable of discerning signals that travel different paths because a RAKE receiver is incorporated. In fact, TIA IS-95A does not address equalization of delay spread on the radio link, and thus the mobile station's receiver does not have an equalizer.

Quantification of propagation between floors is important for in-building wireless system of multifloor buildings that need to share frequencies within the building. Frequencies are reused on different floors to avoid co-channel interference. The type of building material, the aspect ratio of the building sides, and the types of windows have been shown to have an impact on the RF attenuation between floors. Measurements have indicated that loss between floors does not increase linearly in dB with increasing separation distance. The greatest floor attenuation in dB occurs when the receiver and transmitter are separated by a single floor. The overall path loss increases at a smaller rate as the numbers of floors increase. Typical values of attentuation between floors is 15 dB for one floor of separation and an additional 6–10 dB per floor separation up to four floors of separation. For five or more floors of separation, the path loss will increase by only a few dB for each additional floor (see table 9.2).

Table 9.2 Mean Path Loss Exponents and Standard Deviations

Type	n	σ, dB
All building		
All locations	3.14	16.3
Same floor	2.76	12.9
Through 1 floor	4.19	5.1
Through 2 floors	5.04	6.5
Office building 1		
Entire building	3.54	12.8
Same floor	3.27	11.2
West wing 5th floor	2.68	8.1
Central wing 5th floor	4.01	4.3
West wing 4th floor	3.18	4.4
Grocery store	1.81	5.2
Retail store	2.18	8.7
Office building 2		
Entire building	4.33	13.3
Same floor	3.25	5.2

The signal strength received inside of a building due to an external transmitter is important for wireless systems that share frequencies with neighboring buildings or with an outdoor system. Experimental studies have shown that the signal strength received inside a building increases with height. At lower floors of a building, the urban cluster induces greater attenuation and reduces the level of penetration. At higher floors, an LOS path may exist, thus causing a stronger incident signal at the exterior wall of the building. RF penetration is found to be a function of frequency as well as height within a building. Penetration loss decreases with increasing frequency. Measurements made in front of a window showed 6 dB less penetration loss on coverage than those measurements made in parts of the building without window. Experimental studies also showed that building penetration loss decreased at a rate of about 2 dB per floor from ground level up to the 10th floor and then began to increase around the 10th floor. The increase in penetration loss at the higher floors was attributed to shadowing effects of adjacent buildings.

The mean path loss is a function of distance to the nth power [8]:

$$L_{50}(R) = L(R_0) + 10 \times n \log\left(\frac{R}{R_0}\right) \text{ dB} \qquad (9.15)$$

where $L_{50}(R)$ = mean path loss (dB),
       $L(R_0)$ = path loss from transmitter to reference distance R_0 (dB),
       n = mean path loss exponent,
       R = distance from the transmitter (m),
       R_0 = reference distance from the transmitter (m).

We choose R_0 equal to 1 m and assume $L(R_0)$ due to free-space propagation from the transmitter to be a 1-m reference distance. Next, we assume that the antenna gain equals the system cable losses[3] and get a path loss $L(R_0)$, of 31.7 dB at 914 MHz over a 1-m free-space path.

The path loss was found to be lognormally distributed about equation (9.15). The mean path loss exponent n and standard deviation σ are the parameters that depend on building type, building wing, and number of floors between the transmitter and receiver. The path loss at a transmitter-receiver (T-R) separation of R meters can be given as

$$L(R) = L_{50}(R) + X_\sigma \text{ dB} \qquad (9.16)$$

where $L(R)$ = path loss at a T-R separation distance R meters,
       X_σ = zero mean lognormally distributed random variable with standard deviation σ dB.

Table 9.2 gives a summary of the mean path loss exponents and standard deviation about the mean for different environments [8].

In the multifloor environment, equation (9.15) can be modified to emphasize the mean path loss exponent as a function of the number of floors between transmitter and receiver. The value of n (multifloor) is given in table 9.2.

$$L_{50}(R) = L(R_0) + 10 \times n(multifloor) \log\left(\frac{R}{R_0}\right) \qquad (9.17)$$

Another path loss prediction model suggested in [8] uses the floor attenuation factor (FAF). A constant floor attenuation factor (in dB), which is a function of the number of floors and building type, was

---

3. This is obviously not always true.

included in the mean path loss predicted by a path loss model that uses the *same floor* path loss exponent for the particular building type.

$$L_{50}(R) = L(R_0) + 10 \times n(same-floor) \log\left(\frac{R}{R_0}\right) + FAF \text{ dB} \quad (9.18)$$

where R is in meters and $L(R_0) = 31.7$ dB at 914 MHz.

Table 9.3 provides the floor attentuation factors and standard deviation (in dB) of the difference between the measured and predicted path loss. Values for the floor attentuation factor in table 9.3 are an average (in dB) of the difference between the path loss observed at multifloor locations and the mean path loss predicted by the simple R^n model [equation (9.15)], where n is the *same floor* exponent listed in table 9.2 for the particular building structure and R is the shortest distance measured in three dimensions between the transmitter and the receiver.

Table 9.3 Average Floor Attenuation Factors

	FAF, dB	σ
Office building 1		
Through 1 floor	12.9	7.0
Through 2 floors	18.7	2.8
Through 3 floors	24.4	1.7
Through 4 floors	27.0	1.5
Office building 2		
Through 1 floor	16.2	2.9
Through 2 floors	27.5	5.4
Through 3 floors	31.6	7.2

Soft Partition and Concrete Wall Attenuation Factor Model. The path loss effects of *soft partitions* and *concrete walls* (in dB) between the transmitter and receiver for the same floor was modeled in [5] and has been given as

$$L_{50}(R) = 20 \log\left(\frac{4\pi R}{\lambda}\right) + p \times AF(soft-partition) + q \quad (9.19)$$
$$\times AF(concrete-wall)$$

where: p = number of soft partitions between the transmitter and the receiver,

q = number of concrete walls between the transmitter and the receiver,

λ = wavelength (m),

AF = 1.39 dB for each soft partition,

AF = 2.38 for each concrete wall.

E x a m p l e   9 . 1

Use the two models [equations (9.17) and (9.18)] to predict the mean path loss at a distance R = 30 m through three floors of an office building. Assume the mean path loss exponent for *same floor* measurements in the building is n = 3.27, the mean path loss exponent for three-floor measurements is n = 5.22, and the average FAF is 24.4 dB.

From equation (9.17),

$$L_{50}(30) = 31.7 + 10 \times 5.22 \log\left(\frac{30}{1}\right) = 108.8 \ \text{dB}$$

From equation (9.18),

$$L_{50}(30) = 31.7 + 10 \times 3.27 \log\left(\frac{30}{1}\right) + 24.4 = 104.4 \ \text{dB}$$

The results obtained by the two models are fairly close.

9.4 LINK BUDGETS

The link budgets are intended to provide necessary calculations for the ratio of the received bit energy (E_b) to the thermal noise (N_0) plus interference density (I_0) based on the transmit power, transmit and receive antenna gains, receiver noise figure, channel capacity factor, and propagation as well as interference environment.

Link budgets are required for the forward and reverse traffic channels, pilot channel, paging channel, and sync channel.

9.4.1 Forward Direction

To calculate the effective signal-to-interference ratio for the forward pilot, sync, paging, and traffic channels, we need to calculate the received signal power and received interference on each channel. The following equations enable us to complete the calculations.

1. Traffic Channel Effective Radiated Power:

$$p_t = \frac{P_t}{N_t C_f} \tag{9.20}$$

or

$$p_t = P_t - 10 \log N_t - 10 \log C_f \quad \text{(dBm)} \tag{9.21}$$

where p_t = the effective radiated power (ERP) of the traffic channel (dBm),
P_t = the ERP of all traffic channels from the transmit antenna of cell site (dBm),
N_t = the number of traffic channels supported by the sector or cell,
C_f = the channel voice activity factor (typical value is 0.4–0.6)

2. Power per User:

$$p_u = p_t - G_t - L_c \quad \text{(dBm)} \tag{9.22}$$

where p_u = the traffic channel power per user (dBm),
G_t = the gain of the cell transmit antenna (dB), and
L_c = the transmit filter and cable loss between the output of the linear amplifier circuit (LAC) and the input of transmit antenna (dB).

3. Total Base Station ERP:

$$P_c = 10 \log [10^{0.1 p_t} + 10^{0.1 p_s} + 10^{0.1 p_p} + 10^{0.1 p_{pg}}] \quad \text{(dBm)} \tag{9.23}$$

where P_c = the total base station ERP (dBm),
p_s = the ERP of the sync channel (dBm),
p_p = the ERP of the pilot channel (dBm),
p_{pg} = the ERP of the paging channel (dBm).

4. Base Station Power Amplifier:

$$P_a = P_c - G_t - L_c \tag{9.24}$$

where P_a = the total power of all traffic channels, pilot, paging, and sync channels at the output of the amplifier.

5. Total Mobile Received Power:

$$p_m = P_c + L_p + A_l + G_m + L_m \quad \text{(dBm)} \tag{9.25}$$

where p_m = the total mobile received power (dBm),

L_p = the average propagation path loss between the base station and the mobile (dB),

A_l = the allowance for lognormal shadow loss due to local terrain for a given coverage probability (dB),

G_m = the (receive) gain of the mobile antenna (dB),

L_m = the mobile receiver cable and connector losses (dB).

6. Received Traffic Channel Power:

$$p_{tr} = p_t + L_p + A_l + G_m + L_m \qquad \text{(dBm)} \qquad (9.26)$$

where p_{tr} = the received power of traffic channel by the mobile from the serving base station.

7. Received Pilot Power:

$$p_{pr} = p_p + L_p + A_l + G_m + L_m \qquad \text{(dBm)} \qquad (9.27)$$

where p_{pr} = the received power of the pilot channel by the mobile from the serving base station.

8. Received Paging Channel Power:

$$p_{pgr} = p_{pg} + L_p + A_l + G_m + L_m \qquad \text{(dBm)} \qquad (9.28)$$

where p_{pgr} = the received power of the paging channel by the mobile from the serving base station.

9. Received Sync Channel Power:

$$p_{sr} = p_s + L_p + A_l + G_m + L_m \qquad \text{(dBm)} \qquad (9.29)$$

where p_{sr} = the received power of the sync channel by the mobile from the serving base station.

10. Interference from Other Users (same Base Station) on the Traffic Channel:

$$I_{ut} = 10 \log[10^{0.1 p_m} - 10^{0.1 p_{tr}}] - 10 \log B_w \qquad (9.30)$$

where I_{ut} = the density of interference from other users on the traffic channel (dBm/Hz),

B_W = the bandwidth (Hz).

11. Interference from Other Base Stations on the Traffic Channel:

$$I_{ct} = I_{ut} + 10 \log\left[\frac{1}{f_r} - 1\right] \qquad (9.31)$$

where I_{ct} = the density of interference from other base stations on the traffic channel (dBm/Hz),

f_r = frequency reuse factor (typical value is 0.65).

12. Interference Density for the Traffic Channel:

$$I_t = 10 \log[10^{0.1 I_{ut}} + 10^{0.1 I_{ct}}] \tag{9.32}$$

where I_t = the density of interference on the traffic channel (dBm/Hz).

13. Interference from Other Users (Same Base Station) to the Pilot Channel:

$$I_{up} = p_m - 10 \log B_w \tag{9.33}$$

where I_{up} = the density of interference from other users on the pilot channel (dBm/Hz).

14. Interference from Other Base Stations on the Pilot Channel:

$$I_{cp} = I_{up} + 10 \log\left[\frac{1}{f_r} - 1\right] \tag{9.34}$$

where I_{cp} = the density of interference from other base stations on the pilot channel (dBm/Hz).

15. Interference Density for the Pilot Channel:

$$I_p = 10 \log[10^{0.1 I_{up}} + 10^{0.1 I_{cp}}] \tag{9.35}$$

where I_p = the density of interference for the pilot channel (dBm/Hz).

16. Interference from Other Users (Same Base Station) to the Paging Channel:

$$I_{upg} = 10 \log[10^{0.1 p_m} - 10^{0.1 p_{pgr}}] - 10 \log B_w \tag{9.36}$$

where I_{upg} = the density of interference from other users on the paging channel (dBm/Hz).

17. Interference from other Base Stations on the Paging Channel:

$$I_{cpg} = I_{upg} + 10 \log\left[\frac{1}{f_r} - 1\right] \tag{9.37}$$

where I_{cpg} = the density of interference from other base stations on the paging channel (dBm/Hz)

18. Interference Density for the Paging Channel:

$$I_{pg} = 10 \log[10^{0.1 I_{upg}} + 10^{0.1 I_{cpg}}] \qquad (9.38)$$

where I_{pg} =density of interference for the paging channel (dBm/Hz)

19. Interference from Other Users (Same Base Station) on the Sync Channel:

$$I_{us} = 10 \log[10^{0.1 p_m} - 10^{0.1 p_{sr}}] - 10 \log B_w \qquad (9.39)$$

where I_{us} = the density of interference from other users on the sync channel (dBm/Hz).

20. Interference from Other Base Stations on the Sync Channel:

$$I_{cs} = I_{us} + 10 \log\left[\frac{1}{f_r} - 1\right] \qquad (9.40)$$

where I_{cs} = the density of interference from other base stations on the sync channel (dBm/Hz).

21. Interference Density for the Sync Channel:

$$I_s = 10 \log[10^{0.1 I_{us}} + 10^{0.1 I_{cs}}] \qquad (9.41)$$

where I_s = the density of interference on the sync channel (dBm/Hz).

22. Thermal Noise:

$$N_0 = 10 \log(290 \times 1.38 \times 10^{-23}) + N_f + 30 \qquad (9.42)$$

where N_0 = thermal noise density (dBm/Hz),
N_f = noise figure of mobile receiver (dB).

23. Traffic Channel Signal-to-Noise Plus Interference Ratio:

$$\frac{E_b}{N_0 + I_t} = p_{tr} - 10 \log b_{rt} - 10 \log[10^{0.1 I_t} + 10^{0.1 N_0}] \qquad (9.43)$$

where b_{rt} = the bit rate of the traffic channel (bps).

24. Pilot Channel Signal-to-Noise Plus Interference Ratio:

$$\frac{E_b}{N_0 + I_p} = p_{pr} - 10 \log B_w - 10 \log[10^{0.1 I_p} + 10^{0.1 N_0}] \qquad (9.44)$$

25. Paging Channel Signal-to-Noise Plus Interference Ratio:

$$\frac{E_b}{N_0 + I_{pg}} = p_{pgr} - 10 \log b_{rpg} - 10 \log[10^{0.1 I_{pg}} + 10^{0.1 N_0}] \qquad (9.45)$$

where b_{rpg} = the bit rate of the paging channel (bps).

26. Sync Channel Signal-to-Noise Plus Interference Ratio:

$$\frac{E_b}{N_0 + I_s} = p_{sr} - 10 \log b_{rs} - 10 \log[10^{0.1 I_s} + 10^{0.1 N_0}] \qquad (9.46)$$

where b_{rs} = the bit rate of the sync channel (bps).

9.4.2 Reverse Direction

The calculations on the reverse channel are similar to calculations on the forward channels; we use the following equations.

1. Mobile Power Amplifier:

$$P_{ma} = P_{me} - L_m - G_m \qquad (9.47)$$

where P_{ma} = the power output of the mobile power amplifier (dBm),[4]
P_{me} = the ERP from the transmit antenna of the
mobile (dBm),
L_m = the transmit filter and cable loss between the output of the power amplifier and input of the
transmit antenna of the mobile (dB),
G_m = the gain of the mobile transmit antenna (dB).

2. Base Station Received Power per User:

$$P_{cu} = P_{me} + L_p + A_l + G_t + L_t \qquad (9.48)$$

where P_{cu} = the received power of a traffic channel from a mobile
by the serving base station (dBm),
L_p = the mean propagation path loss between the
mobile and the base station (dB),

---

4. dBm is referenced to 1 milliwatt.

A_l = the lognormal shadow/fade allowance due to local
    terrain for a given coverage probability (dB),
G_t = the (receive) gain of the base station antenna,
L_t = the base station receiver cable and connector
    losses (dB),

3. Interference Density of Other Mobiles in the Serving Base Station:

$$I_{utr} = P_{cu} + 10 \log(N_t - 1) + 10 \log C_a - 10 \log(B_w) \qquad (9.49)$$

where I_{utr} = the interference density from other mobiles in
    the serving base station (dBm/Hz),
C_a = the channel voice activity factor (typical value
    is 0.4–0.6),
N_t = the number of traffic channels supported per
    sector/base station,
B_W = the bandwidth (Hz).

4. Interference Density from the Mobiles in Other Base Stations:

$$I_{ctr} = I_{utr} + 10 \log\left(\frac{1}{f_r} - 1\right) \qquad (9.50)$$

where I_{ctr} = the density of interference from the mobiles in
    other base stations (dBm/Hz),
f_r = frequency reuse factor (typical value = 0.6).

5. Interference Density from Other Mobiles in the Serving Base Station and Other Base Stations:

$$I_{tr} = 10 \log[10^{0.1 I_{utr}} + 10^{0.1 I_{ctr}}] \qquad (9.51)$$

where I_{tr} = the density of interference density from other
    mobiles in the serving base station and
    from other base stations (dBm/Hz).

6. Thermal Noise Density:

$$N_0 = 10 \log(290 \times 1.38 \times 10^{-23}) + N_f + 30 \qquad (9.52)$$

where N_0 = the thermal noise density at the reference ther-
    mal noise temperature of 290 K,
N_f = the noise figure of the base station receiver (dB).

7. Reverse Traffic Channel Signal-to-Noise Plus Interference Ratio:

$$\frac{E_b}{N_0 + I_{tr}} = P_{cu} - 10 \log b_{rr} - 10 \log[10^{0.1I_{tr}} + 10^{0.1N_0}] \qquad (9.53)$$

where b_{rr} = the bit rate on the reverse traffic channel (bps).

E x a m p l e   9 . 2

Use the following data and perform necessary calculations to develop link budgets in the forward and reverse direction:

Total traffic channels ERP	=	57 dBm
Mobile ERP	=	20 dBm
Numbers of users	=	20
Traffic channel activity factor	=	0.6
Pilot channel ERP	=	51.5 dBm
Paging channel ERP	=	46.94 dBm
Sync channel ERP	=	41.5 dBm
Transmit filter and cable losses	=	−2.5 dB
Mobile receive cable and body losses	=	−3 dB
Cell transmit antenna gain	=	14 dB
Mobile antenna gain	=	0 dB
Mean propagation losses	=	−146 dB
Lognormal shadow/fade allowance	=	−6.2 dB
Base station noise figure	=	5 dB
Mobile noise figure	=	8 dB
Forward traffic channel bit rate	=	9600 bps
Reverse traffic channel bit rate	=	9600 bps
Paging channel bit rate	=	9600 bps
Sync channel bit rate	=	9600 bps
Reverse traffic channel bit rate	=	9600 bps

Forward Direction:

$$p_t = 57 - 10 \log 20 - 10 \log 0.6 = 57 - 13.01 + 2.22 = 46.21 \text{ dBm}$$

$$P_c = 10 \log[10^{5.7} + 10^{5.15} + 10^{4.69} + 10^{4.15}] = 58.49 \text{ dBm}$$

$$p_u = 46.21 - 14 + 2.5 = 34.71 \text{ dBm}$$

$$P_a = 58.49 - 14 + 2.5 = 46.99 \text{ dBm}$$

$$p_m = 58.49 - 146 - 6.2 - 0 - 3 = -96.71 \text{ dBm}$$

$$p_{tr} = 46.21 - 146 - 6.2 - 0 - 3 = -108.99 \text{ dBm}$$

$$p_{pr} = 51.5 - 146 - 6.2 - 0 - 3 = -103.70 \text{ dBm}$$

$$p_{pgr} = 46.94 - 146 - 6.2 - 0 - 3 = -108.26 \text{ dBm}$$

$$p_{sr} = 41.5 - 146 - 6.2 - 0 - 3 = -113.70 \text{ dBm}$$

$$I_{ut} = 10 \log[10^{-9.671} - 10^{-10.899}] - 10 \log(1.2288 \times 10^6) = -157.87 \text{ dBm/Hz}$$

$$I_{ct} = -157.87 + 10 \log\left[\frac{1}{0.65} - 1\right] = -160.56 \text{ dBm/Hz}$$

$$I_t = 10 \log[10^{-15.787} + 10^{-16.056}] = -156 \text{ dBm/Hz}$$

$$I_{up} = -96.71 - 60.895 = -157.605 \text{ dBm/Hz}$$

$$I_{cp} = -157.605 + 10 \log\left[\frac{1}{0.65} - 1\right] = -160.295 \text{ dBm/Hz}$$

$$I_p = 10 \log[10^{-15.7605} + 10^{-16.0295}] = 155.73 \text{ dBm/Hz}$$

$$I_{upg} = 10 \log[10^{-9.671} - 10^{-10.826}] - 10 \log(1.2288 \times 10^6) = -157.92 \text{ dBm/Hz}$$

$$I_{cpg} = -157.92 + 10 \log\left[\frac{1}{0.65} - 1\right] = -160.61 \text{ dBm/Hz}$$

$$I_{pg} = 10 \log[10^{-15.792} + 10^{-16.061}] = -156.05 \text{ dBm/Hz}$$

$$I_{us} = 10 \log[10^{-9.671} - 10^{-11.37}] - 10 \log(1.2288 \times 10^6) = -157.693 \text{ dBm/Hz}$$

$$I_{cs} = -157.693 + 10 \log\left[\frac{1}{0.65} - 1\right] = -160.382 \text{ dBm/Hz}$$

$$I_s = 10 \log[10^{-15.7693} + 10^{-16.0382}] = -155.82 \text{ dBm/Hz}$$

$$N_0 = 10 \log(290 \times 1.38 \times 10^{-23}) + 8 + 30 = -165.98 \text{ dBm/Hz}$$

Traffic Channel:

$$\frac{E_b}{N_0 + I_t} = -108.99 - 10 \log(9600) - 10 \log[10^{-15.6} + 10^{-16.598}] = 6.764 \text{ dB} \text{ [5]}$$

5. Typical E_b/I_0 value for the traffic channel is between 6 and 7 dB.

Pilot Channel:

$$\frac{E_b}{N_0 + I_p} = -103.7 - 60.685 - 10 \log[10^{-15.573} + 10^{-16.598}] = -9.257 \text{ dB}^{6}$$

Paging Channel:

$$\frac{E_b}{N_0 + I_{pg}} = -108.26 - 39.83 - 10 \log[10^{-15.605} + 10^{-16.598}] = 7.54 \text{ dB}$$

Sync Channel:

$$\frac{E_b}{N_0 + I_s} = -113.7 - 30.792 - 10 \log[10^{-15.582} + 10^{-16.598}] = 10.93 \text{ dB}$$

Reverse Direction:

$$P_{ma} = 20 - (-3) - 0 = 23 \text{ dBm}$$

$$P_{cu} = 20 - 146 - 6.2 + 14 - 2.5 = -120.7 \text{ dBm}$$

$$I_{utr} = -120.7 + 10 \log(20 - 1) + 10 \log 0.6 -$$

$$10 \log(1.2288 \times 10^6) = -171.03 \text{ dBm/Hz}$$

$$I_{ctr} = -171.03 + 10 \log\left[\frac{1}{0.6} - 1\right] = -172.79 \text{ dBm/Hz}$$

$$I_{tr} = 10 \log[10^{-17.103} + 10^{-17.279}] = -168.8 \text{ dBm/Hz}$$

$$N_0 = 10 \log(290 \times 1.38 \times 10^{-23}) + 5 + 30 = -168.98 \text{ dBm/Hz}$$

Traffic Channel:

$$\frac{E_b}{N_0 + I_{tr}} = -120.7 - 10 \log 9600 - 10 \log[10^{-16.88} + 10^{-16.898}] = 5.35 \text{ dB}$$

9.5 DUAL-MODE CDMA MOBILES

The nominal CDMA channel requires 41 contiguous analog channels with a 9-channel spacing as a guard between the edge of a CDMA channel and the adjacent analog channels. This implies that 59 contiguous analog channels should be removed from the service to introduce the first

6. The pilot channel does not carry any information and is used only for pilot acquisition. Thus, the corresponding E_b/I_0 for pilot may be less than the Shannon's limit discussed in chapter 2.

CDMA channel, 100 analog channels for the first and second CDMA channel, etc. In an analog system with a reuse factor of 7, three analog channels per sector, per base station, will be removed to add first CDMA channel (i.e., (59/7)/3 ≈ 3) and an additional 2 analog channels per sector for each additional CDMA channel (i.e., (42/7)/3 = 2).

We consider an example where the base station is equipped with 78 channel elements. Four channel elements on each sector are dedicated to the pilot channel, the sync channel, and two paging/access channels, leaving 22 channel elements per sector remaining for traffic channels. We assume that 22 channel elements are permanently assigned to each sector. The real system pools all 66 channel elements so that any channel element can be assigned to any sector; this method is referred to as dynamic channel assignment. *Dynamic channel assignment* improves the base station capacity as it reduces the probability of blocking on a sector since an idle channel element can be used in any sector. Blocking occurs when the interference from an additional base station will reduce the voice quality below the acceptable limit. Blocking is determined by the channel capacity and not by the amount of hardware used. Using the method of permanently assigning channel elements to a sector gives the lower bound on the total traffic carried by the base station. The traffic carried by the same channel elements will be higher in the real system.

We consider the case where a three-sector analog base station (BS) is equipped with 57 channels that are assigned an $N = 7$ reuse pattern. We remove 3 analog channels to introduce one CDMA channel to serve 22 users. We assume no overflow traffic from CDMA to analog channels and 2 percent probability of blocking and model the traffic using Erlang B-statistic. We also assume that each sector can serve 22 simultaneous calls with the same voice quality as an analog channel. Table 9.4 provides the calculated capacity of the analog base station and the combined CDMA and analog base station. It may be noted that the traffic capacity of the combined base station is double the capacity of the analog base station.

Table 9.4 Comparison of BS capacity with Analog and Combined CDMA and Analog Channels

Analog Voice Channels	CDMA Traffic Channels	Analog Traffic per sector, Erlangs	Total Analog Traffic, Erlangs	CDMA Traffic per sector, Erlangs	Total CDMA Traffic, Erlangs	Total BS Traffic, Erlangs
19	0	12.34	37.0	—	—	37.0
16	22	9.83	29.5	14.9	44.7	74.2

E x a m p l e   9 . 3

Estimate the base station capacity as a function of CDMA user penetration for an overlay system with 2 percent blocking for both analog and digital subscribers. Assume $N = 7$ as the reuse pattern for the analog channels and $N = 1$ for CDMA channels. The average traffic per subscriber is 0.02 Erlangs. See table 9.5.

Introducing CDMA in an analog system results in reducing the analog base station capacity by 20 percent but doubling the total capacity of the base station. This means that at least 20 percent of the offered traffic load must come from dual-mode mobiles before activating CDMA. To realize 100 percent capacity increase in the base station, about 60 percent of the offered traffic load in the base station must be from dual-mode mobiles.

9.6 THE TRANSITION FROM AN ANALOG SYSTEM TO A DIGITAL SYSTEM

The transition from the analog system to the CDMA system as offered by the cellular system equipment vendors and envisioned by the cellular system operators generally falls into three basic modes:

- An overlay design of two separate, independent systems: one is analog and the other is CDMA.
- A completely integrated system where CDMA and analog service is offered everywhere.
- A partially integrated system in which CDMA and analog coverage is provided in part of the overall service area, and only analog coverage exists in the remaining service area.

9.6.1 Overlay Design

In the overlay scenario, a new CDMA system is superimposed over the existing analog system. The CDMA overlay could require a one-to-one digital base station for each analog base station, but it may use a smaller number of larger coverage area cells than the existing analog system to reduce the cost of base stations. Figure 9.2 shows a CDMA base station coverage area that is three times larger than the analog base station coverage area. Thus, it requires one third as many base stations to provide CDMA coverage over the service area. Such an overlay design could be operated as two separate independent systems that are operated by different vendors from separate mobile switching centers or as a single system in which a single MSC controls both types of base stations. Handoffs between the CDMA system and analog system are possible, and handoffs from analog system to CDMA system are not

Table 9.5 Capacity of a cell versus CDMA user Penetration

Analog Channels per Base Station, N_a	CDMA Channels per Base Station, N_c	CDMA Users per Base Station, CE[a]	Percent CDMA Traffic	Analog Mobile, Erlangs	CDMA Mobile, Erlangs	Total Erlangs	Analog Subscribers	CDMA Subscribers	Total Subscribers
57	0	0	0	$12.34 \times 3 =$ 37.0	0	37.0	1850	0	1850
48	1	66	60.24	$9.83 \times 3 =$ 29.5	$14.9 \times 3 =$ 44.7	74.2	1475	2235	3710
42	2	132	80.91	$8.2 \times 3 =$ 24.6	$34.68 \times 3 =$ 104.0	128.6	1230	5200	6430
36	3	198	89.3	$6.62 \times 3 =$ 19.9	$55.33 \times 3 =$ 166	185.9	995	8300	9295
30	4	264	93.73	$5.084 \times 3 =$ 15.3	$76.38 \times 3 =$ 229.1	244.4	765	11,455	12,220
24	5	330	96.43	$3.627 \times 3 =$ 10.9	$97.69 \times 3 =$ 293.1	304.0	545	14,655	15,200
18	6	396	98.13	$2.277 \times 3 =$ 6.8	$119.2 \times 3 =$ 357.6	364.4	340	17,880	18,220
12	7	462	99.22	$1.092 \times 3 =$ 3.3	$140.7 \times 3 =$ 422.2	425.5	165	21,110	21,275
6	8	528	99.86	$0.223 \times 3 =$ 0.7	$162.4 \times 3 =$ 487.3	488	35	24,365	24,400
0	9	594	100	0	$184.2 \times 3 =$ 552.5	552.5	0	27,625	27,625

a. CE = channel element.

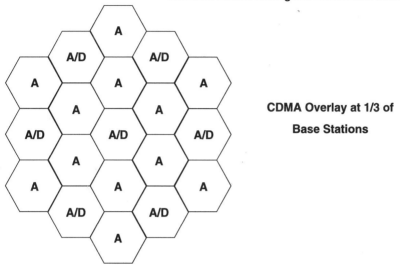

A = Analog Base Stations

A/D = Co-located Analog and CDMA Base Stations

CDMA Overlay at 1/3 of

Base Stations

Figure 9.2 Overlay design.

allowed by the IS-95 standard. The advantages and disadvantages of the full overlay design follow:

Advantages:
- It allows independent analog and CDMA systems in the market.
- If desired, one can use a different vendor for the CDMA system, rather than the analog system vendor.
- The service provider will be able to advertise that CDMA (digital) service is available throughout their service area.
- Since a smaller number of base stations can be used, the investment in digital equipment should be lower than for designs that require digital equipment in a larger number of base stations.

Disadvantages:
- There is capacity loss due to segmentation of the cellular spectrum, since there will be fewer radio frequencies for the analog subscribers.
- The grade of service to analog subscribers, whose numbers may increase if analog terminal "dumping" occurs, will require investing in additional analog base station infrastructure.

- There is increased system operational complexity. The engineering, operation, administration, and maintenance (OA&M) of a two-system CDMA/analog overlay is much more complex than for a single system.
- The "analog-only" base stations may require additional RF filters to reduce the probability that a nearby CDMA mobile will overload the analog base station receiver.

9.6.2 Integrated Design

In this scenario (fig. 9.3), the system is designed to support both analog and CDMA customers everywhere in the service area. As with the other designs, this would be a transitional approach containing the capability for high CDMA capacity in the core and lower CDMA capacity in the noncore areas. Over time, the higher CDMA capacity area would expand to include more and more of the system. The advantages and disadvantages of this approach follow:

Advantages:

- The entire system would have complete digital coverage.
- This scenario avoids receiver overload and other radio problems.
- There is no cellular spectrum segmentation.

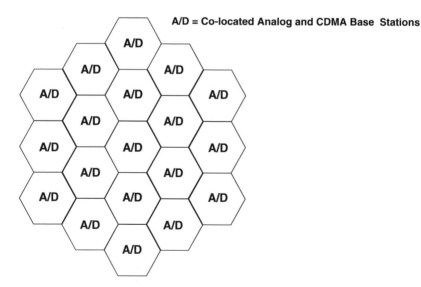

Figure 9.3 Integrated design.

- Full-spectrum efficiency is achieved through high reuse ($N = 1$) everywhere.
- Dual-mode terminals and handoffs between CDMA and analog channels are not required (except for roaming into analog coverage areas).
- OA&M is simplified from that of the overlay system.
- The system operator can advertise digital service everywhere in the service area.

Disadvantages:
- It requires digital equipment everywhere in the system. Although the investment costs can be reduced in areas that need only a low-capacity digital service, it is still a larger investment than for the partial digital system option.

9.6.3 Partial CDMA Coverage, Integrated System

In the partially integrated system, only a part of the system is converted to support analog and digital traffic (usually the core of the system most in need of traffic relief from the digital design). Surrounding the core area of the system, a transition or buffer zone is required to avoid interference between co-channel analog and digital channels. Co-channels are not allowed in the buffer zone. Beyond the buffer zone, base stations can be assigned analog channels that are co-channel with digital channels within the core. A dual-mode mobile that is assigned a digital channel in the core would be handed off to an analog channel as the mobile approaches the edge of the digital coverage area, and then it would be handed to a base station in the buffer zone. Mobiles assigned an analog channel in the analog-only base stations would not transition to a digital channel when inside the digital coverage area.

The transition zone can be gradually moved outward from the core, and the digital coverage can be expanded. The rate of expansion can be determined by the mix of terminals in the system, the need for capacity relief, the strategy for moving the customer base to digital terminals, and the economics of the system operator. Figure 9.4 shows an example of a simplified model with a uniform base station size. The buffer zone between the CDMA/analog base stations and analog-only base stations may require two tiers of base stations depending on the propagation and relative sizes of the actual base stations. If the base stations in the buffer zone are at full capacity, then adding CDMA will reduce the capacity of

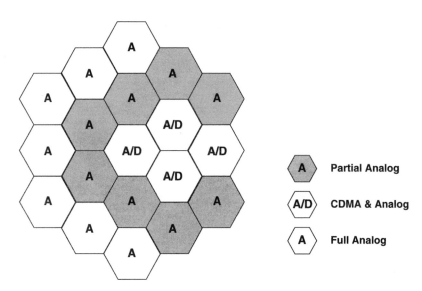

Figure 9.4 Partial CDMA coverage—integrated design.

the buffer zone base stations, since frequencies required by CDMA channels cannot be used in the buffer zone.

Figure 9.5 shows the case where a CDMA base station interferes with the analog mobile. The performance criterion for the analog mobiles is that the carrier signal from the analog base station is 17 dB greater than the total interference on the channel. If the analog system is designed for

$$\frac{S_{anal}}{I_{co-ch}} = 18 \ \text{dB} \tag{9.54}$$

and if

$$\frac{S_{anal}}{I_{co-ch} + I_{CDMA}} \ge 17 \ \text{dB} \tag{9.55}$$

provides adequate voice quality, then

$$\frac{S_{anal}}{I_{CDMA}} > 24 \ \text{dB} \tag{9.56}$$

Table 9.6 lists the relative sizes of CDMA and analog base stations to achieve the criterion as a function of separation between the CDMA and analog center frequencies. With the recommended 9 guard channels,

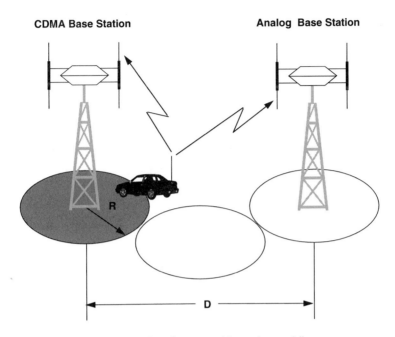

Figure 9.5 CDMA base station interference with analog mobile.

Table 9.6 Required *D/R* Ratios Versus Center Frequency Separation Between CDMA and Analog Center Frequency: CDMA BS Interferes with Analog MS

Center Frequency Separation (f_s), kHz	Required *D/R*
$f_s < 900$	≥ 2.90
$900 \leq f_s < 1980$	≥ 1.33
$f_s \geq 1980$	≥ 1.14

the CDMA base station has to be only slightly farther away than the radius of the analog base station to have adequate analog voice-quality performance. If the analog base station assigns frequencies within the CDMA channel, then the CDMA base station must be about three times the analog base station radius away, or a one base station buffer zone for equal size analog and CDMA base stations. (*R* is the radius of the analog base station, and *D* is the distance between analog and CDMA base stations.)

Figure 9.6 shows a case where the CDMA mobile interferes with the analog base station. The performance criterion for the analog base sta-

CDMA Base Station   **Analog Base Station**

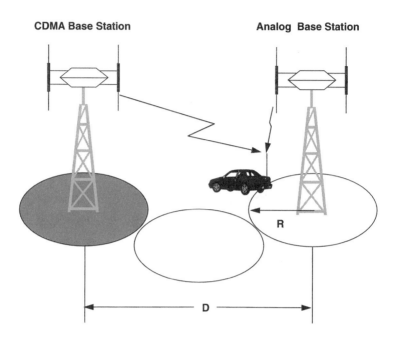

Figure 9.6 CDMA mobile interference with analog base station.

tion is that the received signal must be 17 dB greater than the total received noise plus interference. If the analog system is designed for

$$\frac{S_{anal}}{I_{co-ch}} = 18 \text{ dB} \tag{9.57}$$

and if

$$\frac{S_{anal}}{I_{co-ch} + I_{CDMA}} \geq 17 \text{ dB} \tag{9.58}$$

provides adequate voice quality, then

$$\frac{S_{anal}}{I_{CDMA}} > 24 \text{ dB} \tag{9.59}$$

Table 9.7 lists the required D/R ratios as function of the separation between the CDMA and analog center frequencies to achieve the criterion, where D = distance between the analog and CDMA base station and R = radius of the CDMA base station. With the recommended guard 9 channels, the CDMA mobile can be close to the analog base station before the interference becomes a problem. This situation is less critical than the previous case.

Table 9.7 Required *D/R* Ratio Versus Frequency Separation Between CDMA and Analog Center Frequencies: CDMA MS Interferes with Analog BS

Frequency Separation (f_s), kHz	Required *D/R*
$f_s < 900$	≥ 1.23
$900 \leq f_s < 1980$	≥ 1.05
$f_s \geq 1980$	≥ 1.02

Both rules (tables 9.6 and 9.7) must be satisfied when deploying CDMA base stations to prevent interference to the existing analog base stations. Meeting these criteria will ensure that the interference from the analog transmitters to the CDMA receivers is not a problem as well.

The following are the advantages and disadvantages of the partial integrated design:

Advantages:
- The CDMA capacity advantage can be placed where it is needed most (i.e., in the core system). Only investment for the core system will be needed.
- OA&M is simpler than in the overlay approach.
- The design avoids the receiver overload problem of the overlay design.

Disadvantages:
- The system operator cannot advertise "digital everywhere" service.
- Handoff is required between CDMA and analog coverage areas.
- IS-95 does not provide analog to CDMA handback, so a call initiated in the analog area will not provide any digital features available in the digital coverage area.
- Voice quality changes may be perceived during CDMA to analog handoffs.

E x a m p l e   9 . 4

We consider a small city cellular system that is growing at the predicted growth rate for the next 7 years (see table 9.8). The startup system required 9 omnidirectional coverage base stations and has grown to 29 directional analog base stations to provide service for 36,000 busy hour call attempts (BHCA). Based on the predictions, this system must be expanded to provide capacity for 100,000 BHCA at the end of 7 years. The service provider has chosen to provide CDMA service over the complete coverage area by overlaying the coverage with 10 base stations. The service provider reduces the analog subscribers gradually as indicated in table 9.8. The traffic per subscriber during the busy hour is 0.02 Erlangs.

Table 9.8 Prediction for Analog and Digital Subscribers

End of Year	Analog Subscribers	CDMA Subscribers	Total Subscribers
0	36,000	0	36,000
1	29,000	13,000	42,000
2	24,000	22,000	46,000
3	16,000	39,000	55,000
4	10,000	52,000	62,000
5	4,000	66,000	70,000
6	0	82,000	82,000
7	0	100,000	100,000

Table 9.9 At the End of Year 0

No. of Base Stations		No. of RF Channels/Sector		Capacity, Erlangs/sector		
Analog	CDMA	Analog	CDMA	Analog	CDMA	BHCA
29	—	14	—	8.2	—	36,000

Year 0 (all Analog) (see fig. 9.7 and table 9.9):

- Total traffic during busy hour: $36{,}000 \times 0.02 = 720.0$ Erlangs
- Traffic per sector: $720.0 / (29 \times 3) = 8.276$ Erlangs
- Number of voice channels per sector to provide 2 percent blocking: 14

Year 1 (see fig. 9.8 and table 9.10):

- CDMA traffic: $13{,}000 \times 0.02 = 260$ Erlangs
- Analog traffic: $29{,}000 \times 0.02 = 580.0$ Erlangs

To provide full CDMA coverage, a CDMA minicell is added to 10 base stations.

- CDMA traffic per sector: $260.0 / (10 \times 3) = 8.67$ Erlangs

One CDMA channel provides 14.9 Erlangs per sector with 2 percent system blocking to serve the dual-mode mobiles, so the service provider decides to eliminate all analog channels in the expanded spectrum and uses the spectrum for one CDMA channel in the minicell equipment. This reduces the analog capacity by 3 channels per sector in all 29 base stations.

- Analog traffic per sector: $(29{,}000 \times 0.02) / (29 \times 3) = 6.67$ Erlangs
- Number of analog channel per sector: $14 - 3 = 11$

These 11 channels will carry 6.83 Erlangs of traffic at 4 percent blocking for the analog subscribers. The single CDMA channel will carry all the offered

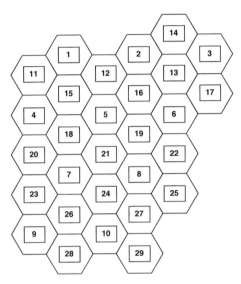

Figure 9.7　Configuration at year 0.

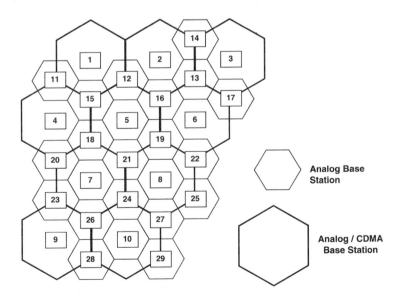

Figure 9.8　Configuration at the end of Year 1 and Year 2.

traffic load with virtually zero blocking for the CDMA subscribers. If the grade of service for the analog subscribers is an unacceptable business strategy, then additional analog capacity would have to be provided by adding channels in the limited spectrum band, possibly resulting in additional co-channel interference. The results are summarized in table 9.10. In the calculations, we assume an average of 22 CDMA calls per sector with 2 percent blocking. As can be seen, the offered CDMA traffic load (8.67 Erlangs per sector) is much lower than the 2 percent blocking capacity (14.9 Erlangs per sector). The CDMA subscribers will experience virtually no blocking, while the analog subscribers will experience 4 percent blocking.

Table 9.10 At the End of Year 1

No. of Base Stations		No. of RF Channels/Sector		Capacity, Erlangs/sector		
Analog	CDMA	Analog	CDMA	Analog	CDMA	BHCA
29	10	11	1	6.83[a]	14.9	42,000

a. 4 percent blocking.

Year 2 (see fig. 9.8 and table 9.11):

- CDMA traffic per sector: $(22{,}000 \times 0.02) / (10 \times 3) = 14.67$ Erlangs

One CDMA channel on 10 base stations with 2 percent blocking provides 14.9 Erlangs per sector.

- Analog traffic per sector: $(24{,}000 \times 0.02) / (29 \times 3) = 5.52$ Erlangs
- Number of analog channels per sector: $14 - 3 = 11$

These 11 channels will carry 5.84 Erlangs of traffic at 2 percent blocking for analog subscribers. Table 9.11 summarizes the results.

Table 9.11 At the End of Year 2

No. of Base Stations		No. of RF Channels/Sector		Capacity, Erlangs/sector		
Analog	CDMA	Analog	CDMA	Analog	CDMA	BHCA
29	10	11	1	5.84	14.9	46,200

Year 3 (see fig. 9.9 and table 9.12):

- CDMA traffic per sector: $(39{,}000 \times 0.02) / (10 \times 3) = 26.0$ Erlangs

Two CDMA channels on 10 base stations with 2 percent blocking gives 34.68 Erlangs per sector.

- Analog traffic per sector: $(16{,}000 \times 0.02) / (23 \times 3) = 4.64$ Erlangs

We remove 6 analog base stations leaving 23 base stations to carry analog traffic.

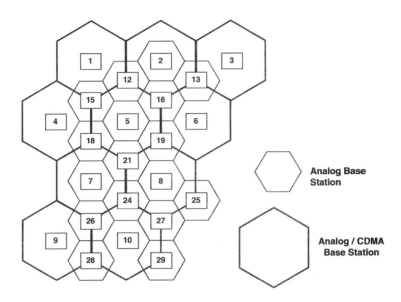

Figure 9.9 Configuration at the end of Year 3.

- Number of analog channels per sector: 14 − 5 = 9

These 9 channels will carry 4.35 Erlangs per sector of traffic at 2 percent blocking for the analog subscribers. Since the offered traffic load is slightly more than the capacity at 2 percent blocking, the analog subscribers will experience about 2.5 percent blocking, whereas the CDMA subscribers will experience almost no blocking since the offered load is less than the capacity. The results are summarized in table 9.12.

Table 9.12 At the End of Year 3

No. of Base Stations		No. of RF Channels/Sector		Capacity, Erlangs/sector		
Analog	CDMA	Analog	CDMA	Analog	CDMA	BHCA
23	10	9	2	4.35	34.68	55,000

Year 4 (see fig. 9.10 and table 9.13):

- CDMA traffic per sector: $(52,000 \times 0.02) / (3 \times 10) = 34.67$ Erlangs

Two CDMA channels on 10 base stations with 2 percent blocking gives 34.68 Erlangs per sector.

- Analog traffic per sector: $(10,000 \times 0.02) / (20 \times 3) = 3.33$ Erlangs
- Number of analog channels per sector: 14 − 5 = 9

We remove 3 analog base stations, leaving 20 base stations to carry analog traffic.

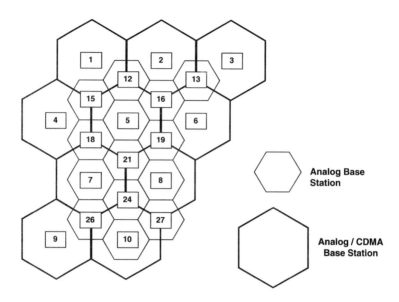

Figure 9.10 Configuration at the end of Year 4.

These 9 channels will carry 4.35 Erlangs per sector of traffic at 2 percent blocking for the analog subscribers. The results are summarized in table 9.13.

Table 9.13 At the End of Year 4

No. of Base Stations		No. of RF Channels/Sector		Capacity, Erlangs/sector		
Analog	CDMA	Analog	CDMA	Analog	CDMA	BHCA
20	10	9	2	4.35	34.68	62,000

Year 5 (see fig. 9.11 and table 9.14):

• CDMA traffic per sector: $(66,000 \times 0.02) / (10 \times 3) = 44.0$ Erlangs

Three CDMA channels on 10 base stations with 2 percent blocking gives 55.33 Erlangs per sector.

• Analog traffic per sector: $(4,000 \times 0.02) / (10 \times 3) = 2.67$ Erlangs

We remove 10 analog base stations, leaving 10 base stations to carry analog traffic.

• Number of analog channels per sector: $14 - 7 = 7$

These 7 channels will carry 2.94 Erlangs per sector at 2 percent blocking. The results are summarized in table 9.14.

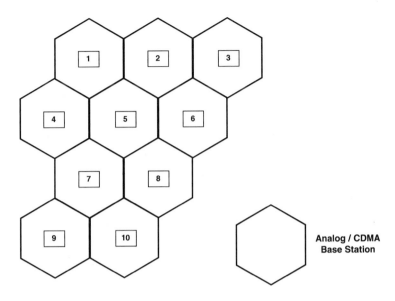

Figure 9.11 Configuration at the end of Year 5.

Table 9.14 At the End of Year 5

No. of Base Stations		No. of RF Channels/Sector		Capacity, Erlangs/sector		
Analog	CDMA	Analog	CDMA	Analog	CDMA	BHCA
10	10	7	3	2.94	55.33	70,000

Year 6 (all digital) (see fig. 9.12 and table 9.15):

- CDMA traffic per sector: $(82,000 \times 0.02) / (10 \times 3) = 54.67$ Erlangs

We eliminate all remaining analog base stations and use 4 CDMA channels on 10 base stations with 2 percent blocking to give 76.38 Erlangs per sector. The results are summarized in table 9.15.

Table 9.15 At the End of Year 6

No. of Base Stations		No. of RF Channels/Sector		Capacity, Erlangs/sector		
Analog	CDMA	Analog	CDMA	Analog	CDMA	BHCA
—	10	—	4	—	76.38	82,000

Year 7 (see fig. 9.12 and table 9.16):

- CDMA traffic per sector: $(100,000 \times 0.02) / (10 \times 3) = 66.67$ Erlangs

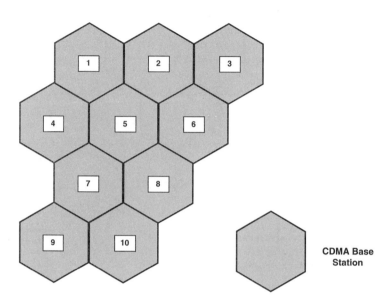

Figure 9.12 Configuration at the end of Year 6 and Year 7.

Four CDMA channels per sector provide 76.38 Erlangs of traffic at 2 percent blocking. The results are given in table 9.16.

Table 9.16 At the End of Year 7

No. of Base Stations		No. of RF Channels/Sector		Capacity, Erlangs/sector		
Analog	CDMA	Analog	CDMA	Analog	CDMA	BHCA
—	10	—	4	—	76.38	100,000

9.7 FACILITIES ENGINEERING

CDMA technology offers a significant capacity improvement with respect to analog and other digital technologies, as was discussed in chapter 2. However, in order to realize these capacity improvements fully, the service provider must properly engineer the CDMA system. This section discusses the engineering of facilities and its relationship to call capacity.

Facilities encompass terrestrial facilities, radio facilities, transcoders (vocoders), and network facilities. As discussed in chapter 4, the transcoders can be physically located at the base station (see fig. 9.13) or at the mobile switching center (see fig. 9.14).

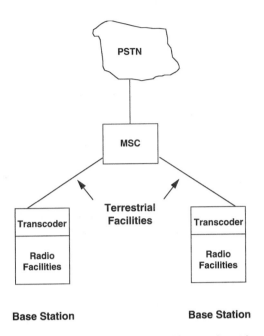

Figure 9.13 Basic CDMA facilities configuration (transcoder at base station).

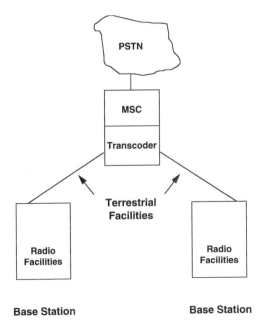

Figure 9.14 Basic CDMA facilities configuration (transcoder at MSC).

Only the radio facility is related to the capacity of the radio link. Also the service provider must configure a sufficient number of CDMA channels to support the maximum number of simultaneous calls. CDMA channels include the pilot channel, sync channel, access channels, paging channels, and traffic channels. Once this is done, the required frequency spectrum can be determined. This section provides an analytical discussion of the capacity of the reverse radio link and a qualitative discussion of the capacity of the forward radio link.

Unlike analog and TDMA technology, CDMA technology does not impose a definite limit on the radio capacity. Rather, CDMA technology exercises a *soft limit*, in which mobile subscribers experience a level of degradation that is related to the total interference and the thermal noise. The capacity is soft since the number of mobile subscribers can be increased if the service provider is willing to lower the grade of service and thus decrease customer satisfaction. With a greater number of simultaneous CDMA calls, the noise floor increases. If the noise floor increases, the probability of receiving a correct frame decreases (i.e., the frame error rate increases). Interference is generated by other CDMA mobile stations that are occupying the same radio spectrum on the same base station or on a different base station. Moreover, for 800-MHz operation, interference is generated if a mobile station that is operating in the analog mode is occupying a portion of the frequency spectrum that is used by the CDMA channel. However, in a properly engineered CDMA system, analog operation is restricted to base stations sufficiently separated from the base station serving the CDMA mobile. This significantly reduces the interference to the CDMA mobile.

In order for a CDMA system to achieve the expected capacity enhancement, it is imperative that power control be properly functioning in both the forward and reverse directions of the radio channel (refer to chapter 5). Power control is executed only on the traffic channels. However, in the following analysis, it is not assumed that power control is perfect. This fact is reflected by assuming that the instantaneous power varies about the desired E_b/I_o level with a lognormal distribution having a standard deviation σ_c. Typical values of σ_c are in the order of 1.5–2.5 dB.

One method of estimating the capacity is to determine the probability that a CDMA channel does not have sufficient bandwidth to accommodate a mobile station for a given frame interval and still satisfy the interference constraints. This event is called an *outage*. During an outage of the reverse radio channel, the frame error rate can exceed the desired maximum limit. This situation is not catastrophic but does lead to degraded service. In the following analysis, the desired interference is given by $(I_0 B_w / N_0 B_w) = (I_0/N_0) < (1/\eta)$. Typically, η is between 0.25 and 0.1, which corresponds to spectral density ratios of (I_0/N_0) between 6 and 10 dB.

The explicit formula for the normalized average user occupancy (λ/μ in terms of Erlangs per sector) is given by [10] and [11]:

$$\frac{\lambda}{\mu} \upsilon(1 + N_I) = K_0 \times F(B, \sigma_c) \qquad (9.60)$$

where λ = average call rate for the entire CDMA system,

$1/\mu$ = average call duration,

υ = voice activity factor,

N_I = interference from base stations outside service area normalized by the interference from the base station serving the service area (in our analysis, N_I will be a constant that is dependent upon the type of CDMA handoff),

$$B = \frac{[Q^{-1}(P_{out})]^2}{K_0},$$

$$K_0 = \frac{\dfrac{B_w}{R}}{\dfrac{E_{b0}}{I_0}} \times (1 - \eta),$$

$$F(B, \sigma_c) = \frac{1}{\alpha_c}\left[1 + \frac{\alpha^3_c \cdot B}{2}\left(1 - \sqrt{1 + \frac{4}{\alpha^3_c \times B}}\right)\right], \text{ in which}$$

$$\alpha_c = e^{(\beta\sigma_c)^2/2}, \text{ and } \beta = (\ln(10))/10 = 0.2303.$$

$Q^{-1}(z)$ is the inverse function of $Q(z)$ where

$$Q(z) = \left(\int_z^{\infty} e^{((-x^2)/2)} dx\right) / (\sqrt{2\pi}).$$

P_{out} is the probability of outage, N_I is the mean interference from neighboring base stations, and E_{b0}/I_0 is the median of the desired E_b/I_0.

E x a m p l e   9 . 5

The following example illustrates the first entry in table 9.18. We calculate the maximum capacity of a CDMA system supporting a bandwidth of 1.25 MHz, 8-kbps vocoders, and 2-way soft handoffs. We assume that the voice activity factor equals 0.4, E_{bo}/I_o = 2.51 or 4 dB, P_{out} = 0.01, the standard deviation of the power control equals 1.78 or 2.5 dB and $(I_o/N_0) \leq 10$. We determine the capacity of the reverse radio link for each antenna sector.

From the above information, we know that η = 0.1. As per TIA IS-95A and IS-96A, the reverse radio link must support 9600 bps to support the 8-kbps vocoder. From equation (9.60), we can calculate K_0:

$$K_0 = \frac{B_w/R}{E_{b0}/I_0} \times \langle 1 - \eta \rangle = \frac{1,250,000/9,600}{2.51} \times (1 - 0.1) = 46.69$$

$$B = \frac{[Q^{-1}(0.01)]^2}{K_0} = \frac{(2.33)^2}{46.69} = 0.116, \quad \alpha_c = e^{(0.23 \times 1.778)^2/2} = 1.0876$$

$$F(B, \sigma_c) = \frac{1}{\alpha_c}\left[1 + \frac{\alpha^3_c \cdot B}{2}\left(1 - \sqrt{1 + \frac{4}{\alpha^3_c \times B}}\right)\right]$$

$$F(B, \sigma_c) = \frac{1}{1.0876} \times \left[1 + \frac{1.0876^3 \times 0.116}{2} \times \left(1 - \sqrt{1 + \frac{4}{1.0876^3 \times 0.116}}\right)\right] = 0.626$$

If a 2-way soft handoff is supported, N_I = 0.77, assuming that propagation losses increase as the fourth power of distance with a standard deviation of 8 dB. If these assumptions are not applicable, then the value of N_I may be different. In such cases, it is recommended that [10] be referenced.

$$\frac{\lambda}{\mu} = \frac{K_0 \times F(B, \sigma_c)}{\upsilon \times (1 + N_I)} = \frac{46.69 \times 0.626}{0.4 \times \langle 1 + 0.77 \rangle} = 41.3 \text{ Erlangs/sector}$$

Using equation (9.60), the following configurations are analyzed. It is assumed that attenuation due to propagation losses decrease as the fourth power of distance and that the lognormal component has a standard deviation of 8 dB. Refer to equation (9.1). The voice activity factor υ is assumed to equal 0.4, and the standard deviation for power control σ_c is assumed to be 2.5 dB. In tables 9.17 to 9.25 we use N_I = 2.38 for hard handoffs, N_I = 0.77 for soft handoffs with a maximum of 2 base stations, and N_I = 0.57 for soft handoffs with a maximum of 3 base stations.

1. Hard handoff only for a CDMA system with a frequency bandwidth of 1.25 MHz and equipped with 8-kbps transcoders (see table 9.17), B_w = 1.25 MHz, R = 9.6 kbps, and $B_w/R = G_p$ = 130;

where G_p is the processing gain, B_w is the bandwidth of the channel, and R is the information rate.

Table 9.17 Configuration 1—Hard Handoffs Only (N_I = 2.38), Bandwidth = 1.25 MHz, 8-kbps Transcoders (B_w/R = 130)

η	E_{b0}/I_0	P_{out}	λ/μ, Erlangs
0.10	2.51 (4 dB)	0.01	21.63
0.25	2.51	0.01	17.38
0.10	3.98 (6 dB)	0.01	12.38
0.25	3.98	0.01	9.86
0.10	2.51	0.05	24.18
0.25	2.51	0.05	19.63
0.10	3.98	0.05	14.23
0.25	3.98	0.05	11.48

2. Soft handoffs with a maximum of two base stations with a 1.25-MHz bandwidth and equipped with 8-kbps transcoders (see table 9.18).

Table 9.18 Configuration 2—Soft Handoffs (N_I = 0.77), Bandwidth = 1.25 MHz, 8-kbps Transcoders (B_w/R = 130)

η	E_{b0}/I_0	P_{out}	λ/μ, Erlangs
0.10	2.51 (4 dB)	0.01	41.30
0.25	2.51	0.01	33.19
0.10	3.98 (6 dB)	0.01	23.63
0.25	3.98	0.01	18.83
0.10	2.51	0.05	46.17
0.25	2.51	0.05	37.49
0.10	3.98	0.05	27.17
0.25	3.98	0.05	21.92

3. Soft handoffs with a maximum of three base stations with a 1.25-MHz bandwidth and equipped with 8-kbps transcoders (see table 9.19).

Table 9.19 Configuration 3—Soft Handoffs (N_l = 0.57), Bandwidth = 1.25 MHz, 8-kbps Transcoders (B_w/R = 130)

η	E_{bo}/I_0	P_{out}	λ/μ, Erlangs
0.10	2.51 (4 dB)	0.01	46.56
0.25	2.51	0.01	37.42
0.10	3.98 (6 dB)	0.01	26.64
0.25	3.98	0.01	21.23
0.10	2.51	0.05	52.05
0.25	2.51	0.05	42.27
0.10	3.98	0.05	30.63
0.25	3.98	0.05	24.71

4. Hard handoffs only with a 1.25-MHz bandwidth and equipped with 13-kbps transcoders (see table 9.20), B_w = 1.25 MHz, R = 14.4 kbps, and B_w/R = 87.

Table 9.20 Configuration 4—Hard Handoffs Only (N_l = 2.38), Bandwidth = 1.25 MHz, 13-kbps Transcoders (B_w/R = 87)

η	E_{bo}/I_0	P_{out}	λ/μ, Erlangs
0.10	2.51 (4 dB)	0.01	13.25
0.25	2.51	0.01	10.46
0.10	3.98 (6 dB)	0.01	7.43
0.25	3.98	0.01	5.86
0.10	2.51	0.05	15.17
0.255	2.51	0.05	12.25
0.10	3.98	0.05	8.79
0.25	3.98	0.05	7.04

5. Soft handoffs with a maximum of two base stations with a 1.25-MHz bandwidth and equipped with 13-kbps transcoders (see table 9.21).

6. Soft handoffs with a maximum of three base stations with a 1.25-MHz bandwidth and equipped with 13-kbps transcoders (see table 9.22).

Table 9.21 Configuration 5—Soft Handoffs (N_I = 0.77), Bandwidth = 1.25 MHz, 13-kbps Transcoders (B_w/R = 87)

η	E_{b0}/I_0	P_{out}	λ/μ, Erlangs
0.10	2.51 (4 dB)	0.01	25.29
0.25	2.51	0.01	20.17
0.10	3.98 (6 dB)	0.01	14.18
0.25	3.98	0.01	11.19
0.10	2.51	0.05	28.97
0.255	2.51	0.05	23.39
0.10	3.98	0.05	16.79
0.25	3.98	0.05	13.45

Table 9.22 Configuration 6—Soft Handoffs (N_I = 0,57), Bandwidth = 1.25 MHz, 13-kbps Transcoders (B_w/R = 87)

η	E_{b0}/I_0	P_{out}	λ/μ, Erlangs
0.10	2.51 (4 dB)	0.01	28.52
0.25	2.51	0.01	22.74
0.10	3.98 (6 dB)	0.01	15.99
0.25	3.98	0.01	12.62
0.10	2.51	0.05	32.66
0.255	2.51	0.05	26.37
0.10	3.98	0.05	18.93
0.25	3.98	0.05	15.17

7. Hard handoffs only with a 10-MHz bandwidth and equipped with 8-kbps transcoders (see table 9.23) B_w = 10 MHz, R = 9.6 kbps, and B_w/R = 1042.

8. Soft handoff with a maximum of two base stations with a bandwidth of 10 MHz and equipped with 8-kbps transcoders (see table 9.24).

9. Soft handoffs with a maximum of three base stations with a 10-MHz bandwidth and equipped with 8-kbps transcoders (see table 9.25).

Table 9.23 Configuration 7—Hard Handoffs Only (N_I = 2.38), Bandwidth = 10 MHz, 8-kbps Transcoders (B_w/R = 1042)

η	E_{b0}/I_0	P_{out}	λ/μ, Erlangs
0.10	2.51 (4 dB)	0.01	221.51
0.25	2.51	0.01	182.21
0.10	3.98 (6 dB)	0.01	134.93
0.25	3.98	0.01	110.62
0.10	2.51	0.05	230.52
0.255	2.51	0.05	190.34
0.10	3.98	0.05	141.87
0.25	3.98	0.05	116.86

Table 9.24 Configuration 8—Soft Handoffs (N_I = 0.77), Bandwidth = 10 MHz, 8-kbps Transcoders (B_w/R = 1042)

η	E_{b0}/I_0	P_{out}	λ/μ, Erlangs
0.10	2.51 (4 dB)	0.01	423.00
0.25	2.51	0.01	347.93
0.10	3.98 (6 dB)	0.01	257.66
0.25	3.98	0.01	211.24
0.10	2.51	0.05	440.21
0.255	2.51	0.05	363.48
0.10	3.98	0.05	270.91
0.25	3.98	0.05	223.16

Table 9.25 Configuration 9—Soft Handoffs (N_I = 0.57), Bandwidth = 10 MHz, 8-kbps Transcoders (B_w/R = 1042)

η	E_{b0}/I_0	P_{out}	λ/μ, Erlangs
0.10	2.51 (4 dB)	0.01	476.88
0.25	2.51	0.01	392.28
0.10	3.98 (6 dB)	0.01	290.48
0.25	3.98	0.01	238.15
0.10	2.51	0.05	496.29
0.255	2.51	0.05	409.79
0.10	3.98	0.05	305.42
0.25	3.98	0.05	251.59

Configurations 1, 2, and 3 correspond to wireless systems supporting standards TIA IS-95A and TIA IS-96A; configurations 4, 5, and 6 correspond to wireless systems supporting standards TIA IS-95A and
Qualcomm's proprietary 13-kbps vocoder; configurations 7, 8, and 9 correspond to TIA SP-2977 and TIA IS-96A.

Comparing tables 9.17 to 9.25, we can make several observations:

- Hard handoffs reduce the capacity on the reverse radio link by
  approximately 50 percent with respect to two-way soft handoffs
  (table 9.17 vis-à-vis table 9.18).
- If a system is equipped with 13-kbps transcoders rather than 8-
  kbps transcoders, the capacity is reduced approximately 40 percent
  (table 9.19 vis-à-vis table 9.22). This observation is consistent with
  our expectations since the rate is increased from 9.6 to 14.4 kbps
  (50 percent increase).
- If the bandwidth increases from 1.25 to 10 MHz (8 times increase),
  the capacity increases approximately 10 times (table 9.18 vis-à-vis
  table 9.24).

We investigate the effect on the reverse radio link capacity if the
power control is perfect (i.e., $\sigma_c = 0$). In this case, $F(B, \sigma_c)$ in equation
(9.60) simplifies to:

$$F(B, \sigma_c) = 1 + \frac{B}{2}\left(1 - \sqrt{1 + \frac{4}{B}}\right) \qquad (9.61)$$

As an example of applying equation (9.61), we can compare the
results in table 9.18 in which $\eta = 0.10$, $(E_{b0}/I_0) = 6$dB, and $P_{out} = 0.01$. In
this case, $(\lambda/\mu) = 23.63$ Erlangs. If the power control is perfect (i.e., $\sigma_c =$
0), (λ/μ) increases to 27.14 Erlangs or a 15 percent increase. However,
equation (9.61) is based upon equation (9.1), which may not adequately
model all radio environments. In such cases, the engineer needs to use
either a more complicated mathematical model or to execute a computer
simulation.

When determining the capacity of the reverse radio link, one must
also include the capacity needed to support the access channels. Both call
setup and registration messages are transmitted on the access channel.
The decrease of the capacity on the reverse radio link is small due to supporting the access channels, typically about 1 percent reduction of the
supportable Erlang traffic. Thus, this reduction is ignored in the determination of the radio link capacity.

As discussed in chapter 7, a call may be simultaneously supported by multiple base stations on the traffic channel. This configuration is called a *soft handoff.* A call may be simultaneously supported by multiple sectors on the same base station. This configuration is called a *softer handoff.* Moreover, a call may be in both a soft handoff and a softer handoff at a particular instance of time. This configuration is called a *soft-softer handoff* (see fig. 9.15). In this example, the mobile station is communicating with sectors D and F (softer handoff with the first base station) and with sector B (in the second base station). Thus, the mobile station is also in soft handoff with both base stations.

When designing a CDMA system for a given call load, the engineer must determine the number of CDMA channel modem (CM) circuits that must be supported at each base station. A CM operates at baseband rather than at RF frequencies. It demodulates the CDMA signal for a given mobile station and combines signals from multiple sectors of a given base station during softer handoffs. The number of CMs is affected by the number of simultaneous soft handoffs but not upon softer handoffs. In the analysis, it is assumed that a CM is not dedicated to a particular sector or CDMA carrier, although this assumption is dependent upon the actual manufacturer's implementation.

The exact distribution of two-way soft handoffs, three-way soft handoffs, softer handoffs, and soft-softer handoffs is very dependent upon the radio environment and radio configuration. The service provider needs to tune the system in order to optimize the radio link capacity. This tuning process includes adjusting the transmitted power of the

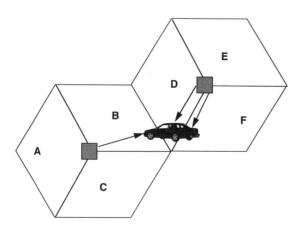

Figure 9.15 Soft-softer handoff configuration.

pilot channels and the threshold levels that trigger handoffs. TIA IS-95A defines several thresholds: T_ADD, T_DROP T_TDROP, and T_COMP. The base station sends the values of these thresholds in the *systems parameter message* (which is transmitted on the paging channel) and the *handoff direction message* or the *extended handoff direction message* (which is transmitted on the traffic channel). During a call, the mobile station measures the strength of the pilot channel of the serving base stations (sectors) and potential candidates. The mobile station sends a *pilot strength measurement message* to the base station in order to initiate a possible handoff for one of the following reasons:

- Pilot's signal strength of a serving cell drops below T_DROP for a duration equal to T_TDROP.[7] T_TDROP is determined by the service provider and ranges from 0 to 319 seconds.
- Pilot's signal strength of a candidate exceeds T_ADD.[8]
- Pilot's signal strength of a candidate exceeds that of a serving base station by T_COMP.[9]

As an example, table 9.26 shows the handoff distribution exhibited during CDMA trials in San Diego, California.

Table 9.26 Handoff Distribution—CDMA Formal Field Test, November 18–23, 1991

Handoff Type	Mean, percent	Standard Deviation, percent
Softer	62.80	19.14
Two-way soft	6.52	19.05
Soft-softer	4.42	7.90
Three-way soft	0.40	0.82
Three-way softer	0.20	0.42
No handoff	25.66	10.80

As a general rule, a CDMA system is tuned so that calls are in some form of soft handoff (i.e., two-way soft, soft-softer, or three-way soft) for

7. As defined in Section 6.6.6.2.5.2 of TIA IS-95A, this pilot is contained in the Active Set.

8. TIA IS-95A refers to the pilot as being in the Neighbor Set or in the Remaining Neighbor Set.

9. TIA IS-95A refers to the pilot as being in the Candidate Set.

approximately 30 percent of the time. Percentages greater than this often are not justified by the improvement of the call quality. It is interesting to note that the aggregated results of the CDMA Formal Field Test (table 9.26) indicate that the total soft handoff (the sum of two-way soft, three-way soft, and soft-softer) is 12 percent of the time. This observation may be rationalized by the fact that calls were in total softer handoff (the sum of softer, soft-softer, or three-way softer) 67 percent of the time. Also, it should be noted that the standard deviation of the handoff distributions is large, particularly for two-way soft and soft-softer handoffs. There are two reasons for this:

- The handoff distribution varies with the base stations that are serving the call. Base stations are tuned differently in order to optimize the call capacity.
- The handoff distribution varies with the path of the mobile station. Even though the mobile station may be served by the same base station, the terrain within the base station's domain varies sufficiently to significantly affect the handoff distributions.

When a call is in a two-way soft, soft-softer, or three-way softer handoff, two CMs are required to support the call. When a call is in a three-way soft handoff, three CMs are needed. Only one CM is needed for the call during a softer handoff or when no handoff configuration occurs.

Example   9 . 6

Assume an equal call distribution across all base stations and sectors with the same handoff distribution. The handoff distribution follows: 40 percent softer handoff, 20 percent two-way soft handoff, 10 percent soft-softer handoff, 29 percent no handoff, and 1 percent three-way soft handoff. There are two CDMA carriers, each having a bandwidth of 1.25 MHz. The system is equipped with transcoders that conform to TIA IS-96A (i.e., 8-kbps voice coding and 9.6 kbps on the physical layer). Assume $\eta = 0.25$, $E_{b0}/I_0 = 6$ dB, and $P_{out} = 0.01$. The system is configured with 10 base stations, each having three sectors. The average call duration is 90 seconds, and each mobile subscriber generates 0.03 Erlang during the busy hour.

Determine the number of calls that can be supported per hour by the system. Determine the number of mobile subscribers that can be provided service if 2 percent blocking is acceptable. Also, determine the number of CMs that must be equipped to support the calculated number of subscribers.

From table 9.19, the capacity of the reverse radio link per sector is:

$$21.23 \ \frac{\text{Erlangs}}{\text{carriers}} \times 2 \text{ carriers } = 42.46 \text{ Erlangs}$$

Thus, each sector can simultaneously support 42 CDMA channels. This does not equal the number of simultaneous calls because some of the channels are assigned as the second and third channels for calls in softer and soft handoff. A CDMA channel is required for each sector configured in the call. Refer to figure 9.16.

In order to determine the number of simultaneous calls that can be supported by the CDMA system, we must associate each call with one sector, even though the given call is being served by multiple sectors. In this example, we assume that the call is associated with the oldest serving sector, although other assignments may be assumed. If complete homogeneity is assumed, we can determine the total number of simultaneous calls supported by the entire system by multiplying the number of sectors ($10 \times 3 = 30$) with the number of simultaneous calls supported by each sector. To illustrate this point, let us determine the number of simultaneous calls supported by sector B as shown in figure 9.16. Softer handoffs are served by sectors B-C and B-A; two-way soft handoffs, by B-G and B-F; three-way soft handoffs, by B-G-F; soft-softer handoffs, by B-C-G, B-C-J, B-C-F, B-C-N, B-A-U, B-A-F, B-A-Q, and B-A-G; no handoffs, by B. Let the number of simultaneous calls supported by the sector be x. Then x can be determined by

$$x + 0.40x + 0.2x + 0.1x + 0.01x = 42$$

where 40 percent of the channels are supporting other sectors in softer handoff, 20 percent of the channels are supporting other base stations in two-way soft handoff, 10 percent of the channels are supporting other base stations in soft-softer handoff, and 1 percent of the channels are supporting base stations in three-way soft handoff.

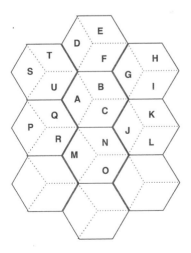

Figure 9.16 Sector configuration for Example 9.6

Thus, x equals 24 simultaneous calls per sector, and the entire CDMA system can serve:

$$10 \text{ cells } \times 3 \frac{\text{sectors}}{\text{base station}} \times 24 \frac{\text{calls}}{\text{sector}} = 720 \text{ calls}$$

In order to determine the number of subscribers that can be supported by this system by each sector during the busy hour, we use the Erlang-B formula. Each sector can support 24 simultaneous calls, which is equivalent to trunks or radios. From Erlang-B tables, we find that 16.63 Erlangs per sector can be supported during the busy hour. Thus, the number of mobile subscribers that the system supports is determined by

$$0.03 \frac{\text{Erlangs}}{\text{subscriber}} \times N(\text{subscribers}) = 16.63 \frac{\text{Erlangs}}{\text{sector}} \times 3 \frac{\text{sectors}}{\text{base station}} \times 10 \text{ base station}$$

$$N = \frac{16.63 \times 3 \times 10}{0.03} \text{ subscribers}$$

$$N = 16630 \text{ subscribers}$$

Next, we calculate the number of CMs that need to be equipped at each sector. We have already determined that each sector can support 42 CDMA channels. If a call is in a softer or no handoff, one CM is needed; for two-way soft and soft-softer handoff, two CMs are needed; for a three-way soft handoff, three CMs are needed. The number of CMs that need to be equipped at each sector for supporting the traffic channels is

$$y + 0.2y + 0.1y + 0.01y = 42 \text{ CM}$$

where 20 percent of the CMs are supporting other base stations in soft handoff, 10 percent of the CMs are supporting other base stations in soft-softer handoff, and 1 percent of the CMs are supporting other base stations in three-way soft handoff.

$$y = 32 \text{ CM}$$

In addition, CMs must be equipped for the access channel. Even though the access channel has a negligible effect upon the reverse radio channel ($16.31 \times 0.01 = 0.02$ Erlang), a CM must be equipped to support this channel.

9.8 CAPACITY OF FORWARD RADIO CHANNEL

We have assumed that the reverse radio channel is the limiting factor when determining the capacity of the radio channel. Qualitatively, we can justify this assumption by the following observations:

- As per TIA IS-95A, the forward radio channel is orthogonal, while the reverse radio channel is approximately orthogonal.
- Interference is from a few sources (base stations) to many receivers (mobile stations) on the forward radio channel, while the interfer-

ence is from many sources (mobile stations) to a few receivers (base stations) on the reverse radio channel.

- The pilot channel supports synchronization on the forward radio channel.

9.9 DESIGN CONSIDERATIONS AT THE BOUNDARY OF A CDMA SYSTEM

Standards do not currently support soft handoffs between CDMA systems that are operated by different service providers. Thus, if a mobile station moves between these CDMA systems, a hard handoff will occur. This results in the reduction of capacity. Comparing table 9.17 with table 9.18, we see that the capacity is degraded by approximately 50 percent. The service provider has several options in order to mitigate this problem. First, the boundary base stations (sectors) can be located at low traffic areas. If this is not practical, additional spectra can be allocated at the boundary sectors. Neither option is very appealing to the service provider. Consequently, TIA TR-45 is currently developing an intervendor soft handoff.

9.10 SUMMARY

This chapter discusses principles for engineering a CDMA system including propagation models, link budgets, and facilities engineering. However, extensive RF measurements and RF modeling are needed to plan a real commercial system. The intent of this chapter is to provide some tools for a better understanding of achieving this goal.

9.11 REFERENCES

1. Hata, M., "Empirical Formula for Propagation Loss in Mobile Radio Services," *IEEE Trans., Veh., Technol.*, VT-29, 1980, pp. 317–325.
2. ITT, *Reference Data for Radio Engineers*, Fifth Edition, Howard W. Sams and Company, Indianapolis, 1969.
3. Jakes, W., *Microwave Mobile Communications*, John Wiley & Sons, New York, 1974.
4. Lee, William, C. Y., *Mobile Cellular Telecommunications System*, McGraw-Hill Book Company, New York, 1989.
5. Motley, A. J., and Keenan, J. M., "Radio Coverage in Buildings," *British Telecom Tech. J.*, *Special Issue on Mobile Communications*, 8, (1), 19–24 (January 1990).
6. Okumura, Y., Ohmori, E., Kawano, T., and Fukuda, K., "Field Strength and its Variability in VHF and UHF Land-mobile radio service," *Rev. Elec. Communication Lab.*, 16, 1968, pp. 825–873.
7. Qualcomm, Inc., "CDMA Formal Field Test for November 18–23, 1991," presented to the Telecommunications Industry Association June 8–12, 1992.

8.   Seidel, S. Y., and Rappaport, T. S., "914 MHz Path Loss Prediction Models for Indoor Wireless Communications in Multifloored Buildings," *IEEE Trans., Antennas & Propagation,* 40 (2), February 1992.

9.   TIA/EIA IS-95A, "Mobile Station—Base Station Compatibility Standard for Dual-Mode Wideband Spread Spectrum Cellular System."

10. Viterbi, A. J., *CDMA Principles of Spread Spectrum Communication,* Addison-Wesley Publishing Company, Reading, MA, 1995.

11. Viterbi, A. M., and Viterbi, A. J., "Erlang Capacity of a Power Controlled CDMA System," *IEEE Journal on Selected Areas in Communications,* 11 (6) August 1993, pp. 892–900.

12. Walfisch, J., and Bertoni, H. L., "A Theoretical Model of UHF Propagation in Urban Environment," *IEEE Trans., Antennas & Propagation.,* to be published.

13. Weissberger, M. A., "An Initial Critical Summary of Models for Predicting the Attenuation of Radio Waves by Trees," ESD-TR-81-101, Electromagnet Compat. Analysis Center, Annapolis, MD, July 1982.

Wireless Data

10.1 INTRODUCTION

In this chapter, we discuss wireless data systems including the wide area wireless data systems and the high-speed wireless local area networks. We present activities for wireless data standards and outline the error control methods used by the standards. We also cover packet radio protocols and their channel efficiency formulas. The contention function of packet radio models the mechanism where mobile stations access the network on the access channel. Packet services are one of the four data services that are supported in CDMA. The other three are asynchronous data, facsimile, and short message services (similar to paging), which are also covered in this chapter.

We discuss the standards for data services that are supported by CDMA cellular/personal communications systems and present highlights of the TIA IS-99, TIA IS-637, and TIA IS-657 standards. We describe the architecture for each of the four data services and the protocol stacks that are supported by the services.

We include both sets of standards (CDMA and non-CDMA) for two reasons:

- The WLANs all use some form of spread spectrum communications, either frequency hopping or direct sequence spreading.
- The two methods (WLANs and CDMA) are part of a larger wireless network that many companies are constructing.

With the phenomenal growth of laptop personal computers and the Internet, wireless data are no longer limited to just e-mail or faxes. It encompasses the ability to send and receive data anytime, from any place in the world, and to provide to a user, at a remote location, full access to all of their desktop services that would normally be available at their office PC. Data services are delivering the same promise that voice services have recently delivered: anytime, anywhere communications.

10.2 WIRELESS DATA SYSTEMS

We can classify wireless data systems into two basic categories: wide-area wireless data systems and high-speed WLANs. WLANs and wide area wireless data systems serve different categories of user applications and, therefore, have different system design objectives. Wireless data services are used for transaction processing and for interactive, broadcast, and multicast services. Transaction processing is used for credit card verification, paging, taxi calls, vehicle theft reporting, and notice of voice or electronic mail. Interactive services include database access and remote LAN access. Broadcast services are general information services, weather and traffic advisory services, and advertising. Multicast services are similar to subscribed information services, law enforcement communications, and private bulletin boards.

In the following sections, we briefly describe wide area wireless data systems and WLANs that have been deployed in the United States.

10.2.1 Wide Area Wireless Data Systems

Wide area wireless data systems are designed to provide high mobility, wide area coverage, and low data rate digital data communications to both vehicles and pedestrians. The technical challenge is to design a system that efficiently uses the available bandwidth to serve large numbers of users distributed over wide geographical areas. Table 10.1 gives the details of the wide area wireless packet data systems that have been deployed in the United States. Specialized mobile radio services (SMRS) allocations are centered around 450 MHz and 900 MHz in the United States.

The ARDIS data network was developed by Motorola as a joint venture between Motorola and IBM to support IBM field service repair people. It is now a public service offering and is solely owned by Motorola. RAM Mobile Data is another public offering that uses the Ericsson Mobitex technology. Both the ARDIS and RAM networks are evolving to data

Table 10.1 Wide-Area Wireless Packet Data Systems

	RAM Mobile (Mobitex)	ARDIS (KDT)	Metricom (MDN)	CDPD
Data Rate	19.2 kbps	19.2 kbps	76 kbps	19.2 kbps
Channel Spacing	12.5 kHz	25 kHz	160 kHz	30 kHz
Access	Slotted ALOHA CSMA	CSMA/CD	FHSS (ISM)	Unused AMPS channels
Frequency, MHz	$f_c \sim 900$	$f_c \sim 800$	$f_c \sim 915$	$f_c \sim 800$
Transmit Power, W	0.16–10 under power control	40	1	1.6
Modulation	GMSK	GMSK	GMSK	GMSK, BT = 0.5

rates of 19.2 kbps. They have been designed to use standard, two-way voice, land mobile-radio channels, with 12.5- or 25-kHz channel spacing.

The CDPD technology shares the 30-kHz spaced, 800-MHz voice channels used by the AMPS systems. The data rate is 19.2 kbps. The CDPD base station equipments share cell sites with the voice cellular radio system. The aim is to reduce the cost of providing packet data service by sharing the resources with the voice cellular systems. This strategy is similar to one that has been used by nationwide fixed wireline packet data networks to provide an economically viable data service by using a small portion of the capacity of the networks designed mainly for voice traffic.

Another approach in wide area wireless packet data networks is based on the microcell concept to provide coverage in smaller areas. The microcell data networks are designed for stationary or low-speed users. The basic aim is to reduce the cost of providing wireless data service by using small and inexpensive base stations that can be installed on the utility poles, the sides of buildings, and inside buildings. The strategy is similar to the one being proposed for personal communications networks. Base station-to-base station wireless links are used to reduce the cost of interconnecting a data network. A large microcell network of small inexpensive base stations has been installed in the lower San Francisco Bay Area by Metricom. The slow frequency hopping-spread spectrum in 902- to 928-MHz U.S. industrial scientific medical (ISM) band is used where transmitter power is 1 W maximum. Power control is used to minimize interference and maximize battery lifetime.

10.2.2 High-Speed Wireless Local Area Networks

A WLAN typically supports a limited number of users in a well-defined indoor area. System aspects such as bandwidth efficiency and product standardization are not crucial. The maximum achievable data rate is an important consideration in the selection of a WLAN. The transmission channel characteristics and signal-processing techniques are important.

WLANs are used to extend wired LANs for convenience and mobility. Three different approaches for connectivity of WLANs have been used. The first approach includes the access to the wide area networks (WANs) and metropolitan area networks (MANs). In the wide area, the network transmission systems use the cellular arrangement and the wired long-distance network. The data are packetized to meet the immediate demands of the users' community. A proper form and format of the data are required to prevent excessive overhead and consequent latency in transport. The second approach deals with localized communications services for the added convenience of connections between the building floors and desktop in a dynamic environment. Flexibility to provide quick connections for moves, adds, and changes gives the organization significant improvement over the basic wired LAN. The third approach is the flexible mobile LAN arrangement. This form of connectivity is becoming important in all walks of life and business communities of interest. As the workforce becomes more mobile, the need to provide untethered connectivity is increasing exponentially.

Two different technical approaches exist with the WLANs. These are based on radio and optical technologies. In the radio-based technology, there are two solutions: the licensed microwave radio frequency range (18–23 GHz) or the unlicensed radio frequency range (902–928 MHz, 2.4–2.4835 GHz, and 5.75–5.825 GHz). In the unlicensed radio frequency, there are two options. The first option uses an FHSS technology, whereas the second option uses a DSSS technology. The 902- to 928-MHz frequency band is an unlicensed ISM band that allows manufacturers to supply products with very limited constraints. Newer products are also emerging that use the 2.4-GHz band. The following are the major limitations of the unlicensed frequency band WLANs:

- The system is restricted to 100-mW output.
- The system must not interfere with other radio frequency equipment in the same area.

• The system must go through an FCC-type acceptance process (in the international sector, this is called homologation or type acceptance, and the frequencies may be different, using either 902- to 928-MHz, 2.4- to 2.4835-GHz, or 5.75- to 5.825-GHz frequencies in various ISM bands).

Spread-Spectrum Radio-Based WLANs   WLANs use spread spectrum techniques to allow flexibility and minimize interference, while not being license-bound. WLANs using FHSS and DSSS approaches with different speeds have been produced. The motivation to use spread spectrum for packet radio systems comes from improved multipath resistance, the ability to coexist with other systems, and the antijamming nature of the code. In an office environment, spread spectrum is a promising choice because it reduces the effects of multipath caused by reflections from the walls and increases the mobility of the terminals within the office environment. The low spectral power density per user of spread spectrum permits an overlay with certain existing systems and reduces the concerns about the health-related issues in high-power transmission. Spread spectrum offers the potential for greater range and higher data rates compared with the optical technology. It improves interception resistance and provides data privacy.

Table 10.2 provides a partial list of WLANs available in the United States.

The following sections describe briefly two WLANs: AT&T WaveLAN based on the FHSS technology and Telesystems advanced radio LAN (ARLAN) based on the DSSS technology [1].

AT&T WaveLAN. This system supports speeds up to 20 Mbps and works with various network operating systems. WaveLAN uses a DS quadrature phase-shift keying multiplexing scheme to transmit across the entire ISM band at high signal rates. Through multiplication of the original narrow band signal with the PN-sequence, the code is spread across several frequencies. WaveLAN offers better security, because the conventional radio receiver cannot decode the signal without knowing the actual spreading pattern. WaveLAN can operate up to 800 feet with a power output of 250 mW. It works in any laptop, notebook, or palmtop PC that is equipped for Personal Computer Memory Card International Association (PCMCIA) card. WaveLAN allows users to operate in a cellular network for LANs. Each WaveLAN is assigned its own identification code and can receive data only if its code corresponds to that of the base

Table 10.2 Partial List of WLAN Products

Product	Frequency	Link Rate	User Rate	Protocol	Access	No. of Channel or Spread Factor	Mod/Coding	Power, mW	Network Topology
Altair Plus Motorola	18–19 GHz	15 Mbps	5.7 Mbps	Ethernet	—	—	4-level FSK	25 peak	8 devices per radio
WaveLAN AT&T	902–928 MHz	2 Mbps	1.6 Mbps	Ethernet-like	DSSS	—	DQPSK	250	Peer-to-peer
AirLAN Solectek	902–928 MHz	—	2 Mbps	Ethernet	DSSS	—	DQPSK	250	PCM-CIA with antenna
Freeport Windata	902–928 MHz	16 Mbps	5.7 Mbps	Ethernet	DSSS	32 chips per bit	16 PSK/ trellis	650	Hub
Intersect Persoft, Inc.	902–928 MHz	—	2 Mbps	Ethernet; Token Ring	DSSS	—	DQPSK	250	Hub
LAWN O'Neill Comm.	902–928 MHz	—	38.4 kbps	AX.25	SS	20 users per channel; max. 4 channels	—	20	Peer-to-peer
WiLan WiLan, Inc.	902–928 MHz	20 Mbps	1.5 Mbps per channel	Ethernet; Token Ring	CDMA/ TDMA	3 channels, 10–15 links each	Unconventional	30	Peer-to-peer
Radio Port ALPS Electric	902–928 MHz	—	242 kbps	Ethernet	SS	—	—	100	Peer-to-peer

Table 10.2 Partial List of WLAN Products (Continued)

Product	Frequency	Link Rate	User Rate	Protocol	Access	No. of Channel or Spread Factor	Mod/Coding	Power, mW	Network Topology
ARLAN 600 Telesystem	902–928 MHz, 2.4 GHz	—	1.35 Mbps	Ethernet	FHSS	—	—	1000 Max.	PCs with antenna
Radio Link Cal. Microwave	902–928 MHz, 2.4 GHz	250 kbps	64 kbps	—	FHSS	250 ms/hop 500 kHz space	—	—	Hub
RangeLAN Proxim, Inc.	902–928 MHz	—	242 kbps	Ethernet; Token Ring	DSSS	3 channel	—	100	—
RangeLAN 2 Proxim, Inc.	2.4 GHz	1.6 Mbps	50 kbps	Ethernet; Token Ring	FHSS	10 channels @ 5 kbps; 15 sub-channel each	—	100	Peer-to-peer bridge
Netwave Xircom	2.4 GHz	1 Mbps per adopter	—	Ethernet; Token Ring	FHSS	82 1-MHz channel or hops	—	—	Hub
Freelink Cabletron System	2.4 and 5.8 GHz	—	5.7 Mbps	Ethernet	DSSS	32 chips per bit	16 PSK trellis	100	100

station it occupies. Users can move anywhere within their assigned base station and still be able to communicate within the base station. If users need to move between base stations, they must first stop the application running and then reconfigure their address ID to match with the cell they are moving into. With roaming, this is automatic. WaveLAN is capable of interfacing directly to the backbone cable systems at standard LAN cable speeds.

Telesystems Advanced Radio LAN. ARLAN uses DSSS technology. Using a conventional cable system, ARLAN devices, called *access points*, are attached to the cable to allow for a full range of interconnections. A microcell can be configured from the backbone network by setting an access point to act like a wireless repeater. Telesystems microcellular architecture (TMA) allows the network to cover various applications and various-sized facilities. With multiple base station antennas, the network can be extended to create microcells, each with its own operating area and devices. TMA is supported by firmware in each of the ARLAN devices and supports multiple overlapping base stations creating a seamless network within the building. Handoff from base station to base station is a part of the network concept that allows for LAN connectivity of users who need to move freely throughout departments or floors within the building. Using the SS technology, the system can select various center frequencies and allows for the coexistence of multiple devices operating within the same area serving different needs. ARLAN 600 was designed for high-noise industrial applications and uses a spreading ratio of up to 100. It offers a full range of interfaces for asynchronous and synchronous data transfer from terminals and hosts. The system operates in the 915-MHz and 2.4-GHz frequency ranges and uses packet burst duplex transmission capabilities. Access to the ARLAN network is packet-switched carrier-sensed multiple access with collision avoidance (CSMA/CA) (see section 10.4 for details on CSMA). Power output for these devices is up to 1 W for distances up to 500 feet diameter in an office environment and up to 3000 feet diameter in factories or open plan offices indoors. For line of sight building-to-building communications, the system can achieve distances of 6 miles. With microcell architecture, each base station is capable of handling up to 1 Mbps. The ARLAN 655 and 670 are complete wireless network interface cards that are mounted inside a PC, workstation (WS), or other device. They provide the same functionality as a conventional LAN adapter card and can support multiple topologies in conjunction with the network operating systems.

10.3 WLAN STANDARDS

All standards for WLANs employ unlicensed bands. There are two
approaches that can be used to regulate an unlicensed band. One
approach is based on a standard to allow different vendors to communicate with one another using a set of interoperable rules. IEEE 802.11
and ETSI's RES 10, HIPERLAN, follow this approach. In the second
approach, a minimum set of rules or "spectrum etiquette" is established
to allow terminals designed by different vendors to have a fair share of
the available channel frequency–time resources and coexist in the same
band. This approach does not preclude the first approach. This has been
pursued by WINForum. In a coexisting environment, a vendor can be
interoperable with another vendor by using the same protocol and transmission scheme.

The three major standard activities for WLANs are: IEEE 802.11,
WINForum, and HIPERLAN. IEEE 802.11 developed a standard for
DSSS, FHSS, and infrared light technology using the ISM bands as the
radio channel. The HIPERLAN standard is aimed at the 5.2- and 17.1-
GHz bands in the European countries. The WINForum's goal is to obtain
a PCS band for unlicensed data and voice applications and to develop a
spectrum etiquette for them.

10.3.1 IEEE 802.11

IEEE 802.11 addresses the physical and media access (MAC) protocol layers for peer-to-peer and peer-to-centralized communications topologies using DSSS or FHSS over radio or infrared light technology. Both
SS systems operate in the 2.4- to 2.4835-GHz ISM band. This band has
been selected over the 902- to 928-MHz and 5.725- to 5.85-GHz ISM
bands because it is widely available in most countries. In the 2.4- to
2.4835-GHz band, more than 80-MHz bandwidth is available that is suitable for high-speed data communication. Also, implementation in this
band is more cost effective as compared with the implementation in frequencies that are higher. IEEE 802.11 supports DSSS with binary
phase-shift keying and QPSK modulation for data rates of 1 and 2 Mbps,
respectively, as well as FHSS with GFSK modulation and two hopping
patterns with data rates of 1 and 2 Mbps. For DSSS, the band is divided
into five overlapping 26-MHz subbands centered at 2.412, 2.442, 2.470,
2.427, and 2.457 GHz, with the last two overlapping the first three. This
setup provides five orders of frequency selectivity for the user. It is quite
cost effective to improve the transmission reliability in the presence of

interference or severe frequency selective multipath fading. For FHSS, the channel is divided into 79 subbands, each with a 1-MHz bandwidth, and three patterns of 22 hops are user options. A minimum hop rate of 2.5 hops/second is assigned to provide slow frequency hopping, in which each packet is sent in one hop. If a packet is destroyed, the following packet is sent from another hop for which the channel condition would be different. This approach provides a very effective time–frequency diversity and takes advantage of a retransmission scheme to provide a robust transmission. The IEEE 802.11 standard avoids rigid requirements and leaves room for vendors to maneuver in the following areas:

- **Multiple physical media.** FHSS and DSSS radio, as well as infrared light; additional media may be approved in the future.
- **Common MAC layer regardless of physical layer.** All IEEE 802.11-compliant WLANs use CSMA/CA algorithm similar to Ethernet's CSMA/CD MAC layer.
- **Common frame format.** Frames including headers and error protection fields are the same, regardless of whether the attached wired LAN is 802.3 Ethernet or 802.5 token ring; the access point handles conversion of 802.11 frames to wireline frame format.
- **Multiple on-air data rates.** 1 or 2 Mbps, with possibility of higher rates in the future.
- **Power limit.** A maximum power of 1 W (or +30 dBm), as mandated by the FCC; there is no minimum power requirement, which leaves open the possibility of low-power implementations.

The standard defines the basic media and configuration issues, transmission procedures, throughput requirements, and range characteristics for WLAN technology. The standard focuses more on access applications that involve the use of personal digital assistants (PDAs) and portable PCs rather than trunk applications (see figs. 10.1 and 10.2). Trunk applications use wireless as part of the enterprise backbone for transmitting data from building to building, whereas access applications allow users of portable PCs, PDAs, and other wireless devices to tap into corporate LANs from anywhere in an office or on a factory floor.

The radio transmitter in each user end station is always listening for activity on the WLAN. If one end station is transmitting, another will not. The system has a preset time-out to block a user from dominating the network, avoid unnecessary transmission collisions, and allow priority traffic through. This is the function of the CSMA/CA access control

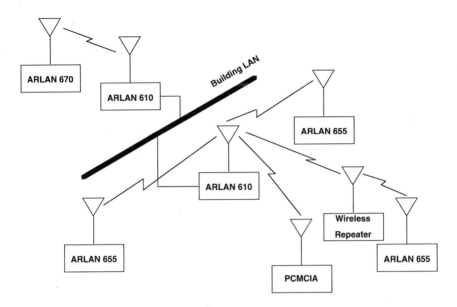

Figure 10.1 Access application for WLAN.

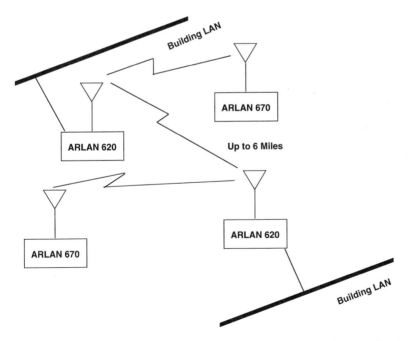

Figure 10.2 Access application for WLAN between two buildings with line of sight.

mechanism. Once it is determined that the network is free, the end station ramps up to full power and sends a preamble (a standard signaling message) to the access point. The *preamble* is a repeated bit pattern followed by a special bit sequence. It allows the access point to lock on to the signal before the data are sent. After the link is established, the end station sends address and protocol information. The header is followed by the data, which are transmitted at the on-air data rate. After the error check word is sent, the end station listens for acknowledgment from the destination. If no acknowledgment is received, the data are resent. The sequence is repeated until all the data have been sent and acknowledged.

The IEEE 802.11 committee has specified that data rates for wireless systems must be either 1 or 2 Mbps. Either the user chooses the rate or the system selects the best one according to the conditions. The on-air data rate includes message headers, retransmissions, and latency (the time between when a network station begins to seek access to a transmission channel and when that access is granted). Header overhead and retransmissions primarily affect performance of large data transfers. Whereas, latency has the greatest effect on short, bursty data transfers, because the latency involved in setting up a transmission introduces more delay than the transmission of message overhead or retransmissions. Therefore, throughput on a WLAN is lower for short messages than for longer messages. The actual throughput of an IEEE 802.11 system on an on-air data rate of 2 Mbps is about 1–1.5 Mbps for long messages and 0.5–1.0 Mbps for short messages. Throughput is also affected by the range of the system. In a typical office environment, the range of an IEEE 802.11 WLAN is 200–300 feet, which is sufficient to cover most partitioned areas and an outside rim of walled offices.

Sensitivity of the system is crucial because signal power can be affected drastically by obstacles. Sensitivity figures are the smallest amounts of received power that the radio can use. The IEEE 802.11 standard requires a sensitivity of less than −80 dBm. One issue that is not addressed by the standard is roaming capability. Roaming is made possible with overlapping WLAN cells in a configuration similar to that used for analog cellular phones. Roaming is considered to be part of the application- or driver-level technology, so vendors will be likely to resort different schemes for achieving it.

10.3.2 Wireless Information Networks Forum

The Wireless Information Networks Forum (WINForum) addresses WLAN and wireless private branch exchange (WPBX) services and focuses on spectrum etiquette to provide fair access to an unlicensed band widely used for different applications and devices. The etiquette does not preclude any common air interface standards or access technologies. It demands listen-before-talk (LBT); thus, a device may not transmit if the spectrum it will occupy is already in use within its range. The power is limited to keep the range short and allow operation in densely populated office areas. The power and connection time are related to the occupied bandwidth to equalize the interference and provide a fair access to frequency–time resources. In the view of WINForum, the asynchronous transmission used in WLAN applications is bursty, begins transmission within milliseconds, uses short bursts that contain large amounts of data, and releases the link quickly. On the other hand, the isochronous transmission, typified by voice services such as a WPBX, uses long holding time, periodic transmission, and flexible link access times that may be extended up to a second. The asynchronous subbands may range from 50 kHz to 10 MHz, whereas the isochronous subbands may be divided into 1.25-MHz segments. The two types are technically contrasting and cannot share the same spectrum.

10.3.3 High-Performance Radio Local Area Network

ETSI's subtechnical committee RES 10 has been assigned the task of developing a standard for the High-Performance Radio Local Area Network (HIPERLAN). The committee secured two bands at 5.12–5.30 GHz and 17.1–17.3 GHz for the HIPERLAN to operate at a minimum useful bit rate of 20 Mbps for point-to-point application with a range of 50 m. It is expected that, at this rate and range, a data rate of 500–1000 Mbps, comparable with fiber distributed data interface (FDDI) for a standard building floor of approximately 1000 m^2, can be achieved. RES 10 is responsible to define a radio transmission technique, including type of modulation, coding, and channel access, as well as the specific protocols.

10.3.4 ARPA

The U.S. Advanced Research Project Agency (ARPA) has sponsored WLAN projects at the University of California at Berkeley (UCB) and University of California at Los Angeles (UCLA). The UCB Infopad project is based on a coordinated network architecture with fixed coordinating nodes and DSSS (CDMA), whereas the UCLA project is for peer-to-peer networks and uses FHSS. Both APRA-sponsored projects are concentrated on the 900-MHz ISM band.

10.4 ACCESS METHODS

10.4.1 Fixed-Assignment Access Methods

In the fixed-assignment access method, a fixed allocation of channel resources (frequency or time or both) is made on a predetermined basis to a single user. The three basic access methods—FDMA, TDMA, and CDMA—are the examples of the fixed-assignment access method. In this section, we discuss only the CDMA method. With CDMA, multiple users operate simultaneously over the entire bandwidth of the time–frequency signal domain, and the signals are kept separate by their distinct user-signal codes. As we discussed in chapter 2, the number of users that can be supported simultaneously by a DS-CDMA system is

$$M = \left[\frac{G_p}{E_b/N_0}\right] \times \frac{1}{1+\beta} \times \alpha \times \frac{1}{\upsilon} \times \lambda \qquad (10.1)$$

where G_p = processing gain = B_w/R_s,
$\quad\quad B_w$ = bandwidth,
$\quad\quad R_s$ = symbol transmission rate,
$\quad\quad E_b/N_0$ = bit energy-to-noise ratio,
$\quad\quad \beta$ = interference factor,
$\quad\quad \alpha$ = power control factor,
$\quad\quad \upsilon$ = voice activity factor (= 1 for data service),
$\quad\quad \lambda$ = gain due to sector antenna.

E x a m p l e   1 0 . 1

We consider a CDMA system that uses QPSK modulation and convolutional coding. The system has a bandwidth of 1.25 MHz and transmits data at 9.6 kbps. Find the number of users that can be supported by the system and the bandwidth efficiency. Assume a three-sector antenna with effective gain = 2.6, $\alpha = 0.9$, and an interference factor $\beta = 0.5$. A bit-error rate of 10^{-3} is required.

$$G_p = \frac{1.25 \times 10^6}{9.6 \times 10^3} = 130.2$$

$$P_b = 10^{-3} = \frac{1}{2}\text{erfc}\sqrt{\frac{E_b}{N_0}}$$

$$\frac{E_b}{N_0} \approx 7 \text{ dB(5)}$$

$$\therefore M = \frac{130.2}{5} \times \frac{1}{1+0.5} \times 2.6 \times 0.9 = 40.6 \approx 40$$

$$\eta_{bw} = \frac{40 \times 9.6}{1.25 \times 10^3} = 0.307 \text{ bit/Hz-sec}$$

Example 10.2

We consider a QPSK/DSSS WLAN that is designed to transmit in the 902- to 928-MHz ISM band. The symbol transmission rate is 0.5 Megasymbols/sec. An orthogonal code with 16 symbols is used. A bit-error rate of 10^{-5} is required. How many users can be supported by the WLAN? A three-sector antenna with gain = 2.6 is used. Assume an interference factor $\beta = 0.5$ to account for the interference from users in other cells, and $\alpha = 0.9$. What is the bandwidth efficiency of the system?

Band width, $B_w = 928 - 902 = 26$ MHz

Data rate, $R = R_s \log_2 16 = 0.5 \log_2 2^4 = 2$ Mbps

$$G_p = \frac{B_w}{R_s} = \frac{26}{0.5} = 52$$

$$P_b = 10^{-5} = \frac{1}{2}\text{erfc}\sqrt{\frac{E_b}{N_0}}$$

$$\frac{E_b}{N_0} \approx 10 \text{ dB}(10)$$

$$\therefore M = \frac{52}{10} \times \frac{1}{1 + 0.5} \times 2.6 \times 0.9 = 8.1 \approx 8$$

$$\eta_{bw} = \frac{8 \times 2}{26} = 0.62 \text{ bit/Hz-sec}$$

10.4.2 Random Access Methods

When each user has a steady flow of information to transmit (e.g., a data file transfer or a facsimile transmission), fixed-assignment access methods are useful because they use communication resources efficiently. However, when the information to be transmitted is bursty in nature, the fixed-assignment access methods result in wasting communication resources. Furthermore, in a cellular system where subscribers are charged based on a channel connection time, the fixed-assignment access methods may be expensive to transmit short messages. Random-access protocols provide flexible and efficient methods for managing a channel access to transmit short messages. The random-access methods give freedom for each user to gain access to the network whenever the user has information to send. Because of this freedom, these schemes result in contention among users accessing the network. Contention may cause collisions and may require retransmission of the information. The commonly used random-access protocols are pure ALOHA, slotted-ALOHA, and CSMA/CD. In the next section we describe briefly details of each of these protocols and provide necessary throughput expressions.

10.4.2.1 Pure ALOHA   In the pure ALOHA scheme, each user transmits information whenever the user has information to send. A user sends information in packets. After sending a packet, the user waits a length of time equal to the round-trip delay for an acknowledgment (ACK) of the packet from the receiver. If no ACK is received, the packet is assumed to be lost in a collision, and it is retransmitted with a randomly selected delay to avoid repeated collisions.[1] The normalized throughput S (average packet arrival rate divided by the maximum throughput) of the pure ALOHA protocol is given as

$$S = Ge^{-2G} \qquad (10.2)$$

where G = normalized offered traffic load.

From equation (10.2), note that the maximum throughput occurs at traffic load $G = 50$ percent and is $S = 1/(2e)$. This is about 0.184. Thus, the best channel utilization with the pure ALOHA protocol is only 18.4 percent.

10.4.2.2 Slotted-ALOHA   In the slotted-ALOHA system, the transmission time is divided into time slots. Each time slot is made exactly equal to packet transmission time. Users are synchronized to the time slots, so that whenever a user has a packet to send, the packet is held and transmitted in the next time slot. With the synchronized time-slots scheme, the interval of a possible collision for any packet is reduced to one packet time from two packet times, as in the pure ALOHA scheme. The normalized throughput S for the slotted-ALOHA protocol is given as

$$S = Ge^{-G} \qquad (10.3)$$

where G = normalized offered traffic load.

---

1. It should be noted that the protocol on CDMA access channels as implemented in TIA IS-95A is based upon the pure ALOHA approach. The mobile station randomizes its attempt for sending a message on the access channel and may retry if an acknowledgment is not received from the base station. For further details, one should reference section 6.6.3.1.1.1 of TIA IS-95A.

The maximum throughput for the slotted-ALOHA occurs at $G = 1.0$ [equation (10.3)], and it is equal to $1/e$ or about 0.368. This implies that at the maximum throughput, 36.8 percent of the time slots carry the successfully transmitted packets.

10.4.2.3 Carrier-Sensed Multiple Access
CSMA protocols have been widely used in both wired and wireless LANs. These protocols provide enhancements over the pure and slotted ALOHA-protocols. The enhancements are achieved through use of additional capability at each user station to sense the transmissions of other user stations. The carrier-sensed information is used to minimize the length of collision intervals. For carrier sensing to be effective, propagation delays must be less than packet transmission times. Two general classes of CSMA protocols are nonpersistent and p-persistent. Each of these classes can be used with the slotted or unslotted operation.

- **Non-persistent CSMA.** A user station does not sense the channel continuously while it is busy. Instead, after sensing the busy condition, it waits for a randomly selected interval of time before sensing again. The algorithm works as follows: If the channel is found to be idle, the packet is transmitted; if the channel is sensed busy, the user station backs off to reschedule the packet to a later time. After backing off, the channel is sensed again, and the algorithm is repeated again.
- **p-persistent CSMA.** The slot length is typically selected to be the maximum propagation delay. When a station has information to transmit, it senses the channel. If the channel is found to be idle, it transmits with probability p. With probability $q = 1 - p$, the user station postpones its action to the next slot, where it senses the channel again. If that slot is idle, the station transmits with probability p or postpones again with probability q. The procedure is repeated until either the frame has been transmitted or the channel is found to be busy. When the channel is detected busy, the station then senses the channel continuously. When it becomes free, it starts the procedure again. If the station initially senses the channel to be busy, it simply waits one slot and applies the preceding procedure.
- **1-persistent CSMA.** 1-Persistent CSMA is the simplest form of the p-persistent CSMA. It signifies the transmission strategy, which is

to transmit with probability 1 as soon as the channel becomes idle. After sending the packet, the user station waits for an ACK. If it is not received within a specified amount of time, the user station waits for a random amount of time and then resumes listening to the channel. When channel is again found to be idle, the packet is retransmitted immediately.

For more details, refer to references [2, 21].

The throughput expressions for the CSMA protocols follow:

- **Unslotted Nonpersistent CSMA**

$$S = \frac{Ge^{-aG}}{G(1 + 2a) + e^{-aG}} \tag{10.4}$$

- **Slotted Nonpersistent CSMA**

$$S = \frac{aGe^{-aG}}{1 - e^{-aG} + a} \tag{10.5}$$

- **Unslotted 1-Persistent CSMA**

$$S = \frac{G[1 + G + aG(1 + G + (aG)/2)]e^{-G(1 + 2a)}}{G(1 + 2a) - (1 - e^{-aG}) + (1 + aG)e^{-G(1 + a)}} \tag{10.6}$$

- **Slotted 1-Persistent CSMA**

$$S = \frac{Ge^{-G(1 + a)}[1 + a - e^{-aG}]}{(1 + a)(1 - e^{-aG}) + ae^{-G(1 + a)}} \tag{10.7}$$

where S = normalized throughput,
       G = normalized offered traffic load,
       a = τ/T_p,
       τ = propagation delay,
       T_p = packet transmission time.

E x a m p l e   1 0 . 3

We consider a WLAN installation in which the maximum propagation delay is 0.4 μs. The WLAN operates at a data rate of 10 Mbps, and each packet has 400 bits. Calculate the throughput with (1) an unslotted nonpersistent, (2) a slotted persistent, and (3) a slotted 1-persistent CSMA protocol.

$$T_p = \frac{400}{10} = 40 \ \mu s$$

$$a = \frac{\tau}{T_p} = \frac{0.4}{40} = 0.01$$

$$G = \frac{40 \times 10^{-6} \times 10 \times 10^{6}}{400} = 1$$

- Slotted Nonpersistent:

$$S = \frac{0.01 \times 1e^{-0.01}}{1 - e^{-0.01} + 0.01} = 0.496$$

- Unslotted Nonpersistent:

$$S = \frac{1 \times e^{-0.01}}{(1 + 0.02) + e^{-0.01}} = 0.493$$

- Slotted 1-Persistent:

$$S = \frac{e^{-1.01}(1 + 0.01 - e^{-0.01})}{(1 + 0.01)(1 - e^{-0.01}) + 0.01e^{-1.01}} = 0.531$$

10.5 ERROR CONTROL SCHEMES

Channel coding and automatic repeat request (ARQ) schemes are used to increase the performance of mobile communication systems. In the physical layer of DS-CDMA system, error detection and correction techniques such as forward error correction (FEC) schemes are used. For some of the data services, higher-layer protocols use ARQ schemes to enable retransmission of any data frames in which an error is detected. The ARQ schemes are classified as follows [21, 22].

Stop and Wait. The sender transmits the first packet numbered 0 after storing a copy of that packet. The sender then waits for an ACK numbered 0, ACK0 of that packet. If the ACK0 does not arrive before a time-out, the sender makes another copy of the first packet, also numbered 0, and transmits it. If the ACK0 arrives before a time-out, the sender discards the copy of the first packet and is ready to transmit the next packet, which it numbers 1. The sender repeats the previous steps, using number 1 instead of 0. The advantages of the Stop and Wait protocol are its simplicity and its small buffer requirements. The sender needs to keep only a copy of the packet that it last transmitted, and the receiver does not need to buffer packets at the data link layer. The main disadvantage of the Stop and Wait protocol is that it does not use the communication link efficiently.

The total time taken to transmit a packet and to prepare for transmitting the next one is

$$T = T_p + 2T_{prop} + 2T_{proc} + T_a \qquad (10.8)$$

The protocol efficiency without error is

$$\eta(0) = \frac{T_p}{T} \qquad (10.9)$$

where T = total time for transmitting a packet,
T_p = transmission time for a packet,
T_{prop} = propagation time of a packet or an ACK,
T_{proc} = processing time for a packet or an ACK,
T_a = transmission time for an ACK.

If p is the probability that a packet or its ACK is corrupted by transmission errors and a successful transmission of a packet and its ACK takes T seconds and occurs with probability $1 - p$, the protocol efficiency for full duplex (FD) and half duplex (HD) operation are given as

$$\eta_{FD} = \frac{(1 - p)T_p}{(1 - p)T + pT_p} \qquad (10.10)$$

$$\eta_{HD} = \frac{(1 - p)T_p}{T} \qquad (10.11)$$

Selective Repeat Protocol (SRP). The data link layer in the receiver delivers exactly one copy of every packet in the correct order. The data link layer in the receiver may get the packets in the wrong order from the physical layer. This occurs, for example, when transmission errors corrupt the first packet and not the second one. The second packet arrives correctly at the receiver before the first. The data link layer in the receiver uses a buffer to store the packets that arrive out of order. Once the data link layer in the receiver has a consecutive group of packets in its buffer, it can deliver them to the network layer. The sender also uses a buffer to store copies of the unacknowledged packets. The number of the packets, which can be held in the sender/receiver buffer is a design parameter. Let W be the number of packets that the sender and receiver buffers can each hold and SRP be the number of packets modulo-$2W$. The protocol efficiency without any error and with an error probability of p is given as

$$\eta(0) = \min\left\{\frac{WT_p}{T}, 1\right\} \tag{10.12}$$

For very large W, the protocol efficiency is

$$\eta(p) = 1 - p \tag{10.13}$$

where T = time-out = WTp

$$\eta(p) = \frac{2 + p(W - 1)}{2 + p(3W - 1)} \tag{10.14}$$

SRP is very efficient, but it requires buffering packets at both the sender and the receiver.

Go-Back-N (GBN). The Go-Back-N protocol allows the sender to have multiple unacknowledged packets without the receiver having to store packets. This is done by not allowing the receiver to accept packets that are out of order. When a time-out timer expires for a packet, the transmitter resends that packet and all subsequent packets. The Go-Back-N protocol improves on the efficiency of the Stop and Wait protocol but is less efficient than SRP. The protocol efficiency is given as

$$\eta_{FD} = \frac{1}{1 + \left(\dfrac{p}{1-p}\right)W} \tag{10.15}$$

Window-Control Operation Based on Reception Memory ARQ. In digital cellular systems, bursty errors occur by multipath fading, shadowing, and handoffs. The bit-error rate fluctuates from 10^{-1} to 10^{-6}. Therefore, the conventional ARQ schemes do not operate well in digital cellular systems. Window-control operation based on reception memory (WORM) ARQ has been suggested for control of dynamic error characteristics. It is a hybrid scheme that combines SRP with GBN. GBN protocol is chosen in the severe error condition, whereas SRP is selected in the normal error condition.

Variable Window and Frame Size GBN and SRP. Since CDMA systems have bursty error characteristics, the error control schemes should have a dynamic adaptation to bursty channel environment. The SRP and GBN with variable window and frame size have been proposed in [22] to improve error control in the CDMA systems. Table 10.3 provides the window and frame size for different bit-error rates. If the error-rate increases, the window and frame size are decreased. In the case of error-rate being small, the window and frame size are increased. The optimum threshold values of bit-error rate (BER) and window and frame sizes were obtained through computer simulation.

Table 10.3 Bit-Error Rate Versus Window and Frame Size

Bit-Error Rate (BER)	Window Size (*W*)	Frame Size, bits
$BER \leq 10^{-4}$	32	172
$10^{-4} < BER < 10^{-3}$	8	80
$10^{-3} < BER < 10^{-2}$	4	40
$10^{-2} < BER$	2	16

In CDMA systems, the forward link consists of pilot, sync, paging, and traffic channels. System information sent on the pilot, sync, and paging channels allows each mobile station to evaluate the BER easily by measuring the ratio of the number of retransmitted frames to the number of transmitted frames over a 2-second period. Thus, the mobile station can change the window and frame size according to the BER.

Example   10.4

We consider a WLAN in which the maximum propagation delay is $4\,\mu s$. The WLAN operates at a data rate of 10 Mbps. The data and ACK packet lengths are 400 and 20 bits, respectively. The processing time for a data or ACK packet is $1\,\mu s$. If the probability p that a data packet or its ACK can be corrupted during transmission is 0.01, find the data link protocol efficiency with (1) Stop and Wait protocol, full duplex, (2) SRP with window size $W = 8$, and (3) Go-Back-N protocol with window size $W = 8$.

$$T_p = \frac{400}{10} = 40\ \mu s$$

$$T_a = \frac{20}{10} = 2\ \mu s$$

$$T_{prop} = 4\ \mu s$$

$$T_{proc} = 1\ \mu s$$

$$T = 40 + 2 \times 4 + 2 \times 1 + 2 = 52\ \mu s$$

Stop and Wait:

$$\eta = \frac{(1 - 0.01) \times 40}{52} = 0.762$$

SRP:

$$\eta = \frac{2 + 0.01(8 - 1)}{2 + 0.01(24 - 1)} = 0.954$$

Go-Back-N:

$$\eta = \frac{1}{1 + 8\left(\dfrac{0.01}{1 - 0.01}\right)} = 0.925$$

10.6 THE DATA SERVICES STANDARD FOR CMDA CELLULAR/ PERSONAL COMMUNICATIONS SYSTEMS

CDMA systems send data using the reference model shown in figure 10.3 as standardized in TIA IS-99 [14]. The following is the description of the reference points:

- Reference point R_m is a physical interface that connects a Terminal Equipment 2 (TE2) to an Mobile Terminal Type 2 (MT2). An MT2 provides a non-ISDN user interface.
- Reference point U_m is a physical interface that connects a Mobile Terminal Type 0 (MT0) or MT2 to a base station. U_m is the air interface. An MT0 is a self-contained data-capable mobile terminal that does not support an external interface.
- Reference point A_i is a physical interface connecting a base station to the PSTN.

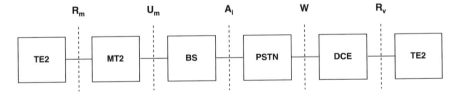

TE2:	Terminal Equipment 2
MT2:	Mobile Termination 2
BS:	Base Station
DCE:	Data Circuit-terminal Equipment
PSTN:	Public Switching Telephone Network

Figure 10.3 Reference model for asynchronous data transmission. (Reproduced under written permission of the copyright holder [TIA].)

- Reference point W is a physical interface that connects data circuit-terminating equipment (DCE) to the PSTN.
- Reference point R_v is a reference point between a TE2 and a Data Communications Equipment (DCE). It may be a physical interface, or it may be internal to the user equipment. CCITT V-Series modems are the examples of TE2; Group-3 FAX is an example of a DCE. TE2 is equivalent to data terminal equipment (DTE) as defined in EIA/TIA-232E [15]; similarly, MT2 is equivalent to DCE.

10.6.1 Asynchronous Data and Group-3 Facsimile

The general approach taken in TIA IS-95A [13] for data services reuses the previously specified physical layer of the IS-95A protocol stack as the physical layer. Figure 10.4 shows the air interface (U_m) protocol stack.

The current TIA standards define three primary services: asynchronous data, Group-3 facsimile, and short message service (SMS). Stan-

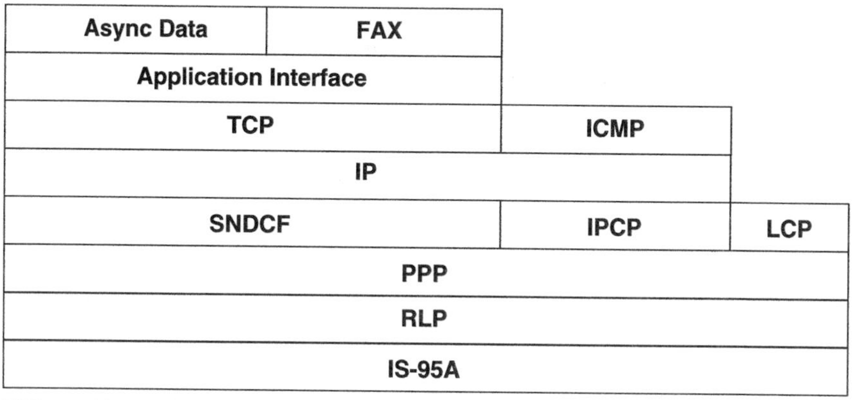

Async Data	FAX	
Application Interface		
TCP	ICMP	
IP		
SNDCF	IPCP	LCP
PPP		
RLP		
IS-95A		

TCP:	Transmission Control Protocol
ICMP:	Internet Control Message Protocol
IP:	Internet Protocol
SNDCF:	Sub-Network Dependent Convergence Function
IPCP:	Internet Protocol Control Protocol
LCP:	Link Control Protocol
PPP:	Point-to-Point Protocol
RLP:	Radio Link Protocol

Figure 10.4 The U_m protocol stack.

dard activities are in progress to define packet data, synchronous data, and other primary services.

IS-95A asynchronous data has been structured as a circuit-switched service in which a dedicated path is established between the data devices for the duration of the call. It is used for connectivity through the PSTN when point-to-point communications to a PC or FAX user is required. For example, for a file transfer involving PC-to-PC communications, the asynchronous data service is the preferred cellular service mode.

The radio link protocol (RLP) employs automatic repeat request, forward error correction, and flow control. Flow control and retransmission of data blocks with errors are used to provide an improved performance in the mobile segment of the data connection at the expense of variations in throughput and delay. Typical raw channel data-error rates for cellular transmission are approximately 10^{-2}. However, an acceptable data transmission usually requires a bit-error rate of about 10^{-6}. In order to achieve this, it requires the design of efficient ARQ and error-correction codes to deal with error characteristics in the mobile environment.

The CDMA protocol stack for data and facsimile (fig. 10.4) has the following layers:

• **The Application Interface Layer** includes an application interface between the data source/destination in the mobile terminal (MT0) or terminal equipment (TE2) and the transport protocol layer. In the base station, the application interface resides between the data source/destination on the network (A_i interface) side and the transport protocol layer. The application interface provides modem control, AT command processing,[2] negotiation of air interface data compression, and data compression over the air interface (optional).

• **The Transport Layer** for CDMA asynchronous data and FAX services is based on Internet transport layer protocol known as transmission control protocol (TCP) [6]. The implementation complies with the requirements for TCP with modifications as described in IS-95 [13]. If the modified procedure is disabled, there is no maxi-

---

2. The AT commands were originally defined by the Hayes Microcomputer Company for their wireline modems. The command set has now been adopted by most wireline and wireless modems. The name AT is derived from the use of AT to preface all commands to the modem.

mum number of retransmission attempts during synchronization, and an established TCP connection remains open until explicitly closed by the mobile station or base station. The application interface sets the value of R_2 in the protocol. The base station follows either the procedure of the Internet control message protocol (ICMP) [7] or the preceding given procedure.

- **The Network Layer** for CDMA async data and FAX services is based on the Internet network layer protocol known as the Internet protocol (IP) [5]. The network layer includes the ICMP [6]. The implementation complies with the requirements of the IP [5] and the requirements for Internet hosts [8] with modifications as described in IS-95 [13]. The interface between the network and transport layer complies with the requirements of the ICMP [7].

- **The Subnetwork Dependent Convergence Function (SNDCF)** performs header compression on the headers of the transport and network layers. This function is negotiated using point-to-point protocol (PPP) and internet protocol control protocol (IPCP) [10]. Mobile stations support Van Jacobson TCP/IP header compression. A minimum of one compression slot is negotiated. Base stations support TCP/IP header compression compatible with that required for mobile stations. Negotiation of the parameters of header compression is carried out using IPCP. The SNDCF sublayer accepts a network layer datagram from the network layer, performs header compression as required, and passes the datagram to the PPP layer, indicating the appropriate PPP identifier. The SNDCF sublayer receives network layer datagrams with compressed or uncompressed headers from the PPP layer, decompresses the datagram header as necessary, and passes the datagram to the network layer.

- **The Data Link Layer** uses PPP [11]. The PPP link control protocol (LCP) is used for initial link establishment and for the negotiation of optional link capabilities. The data link layer uses the PPP and IPCP to negotiate IP addresses and TCP/IP header compression. The data link layer accepts network layer datagrams from the SNDCF and encapsulates them in the PPP information field. The packet is framed using the octet synchronous framing protocol, except that there is no interframe fill. No flag octets are sent between a flag octet that ends one PPP frame and the flag octet that begins the subsequent PPP frame. The framed PPP packets are passed to the RPL layer for transmission. The data link layer accepts received octets from the RLP layer and reassembles the original PPP packets. The PPP process discards any PPP packet for which the received frame check sequence (FCS) is not equal to the computed value.

- **The Internet Protocol Control Protocol Sublayer** supports negotiation of the IP-address (type = 3) and IP-compression protocol (type = 2) parameters. IPCP negotiates a temporary IP address for the mobile station whenever a transport layer connection is actively opened. Mobile stations maintain the temporary IP address only while a transport layer connection is open or being opened; they discard the temporary IP address when the transport layer connection is closed.

- **The Link Control Protocol** layer messages with a protocol identifier of 0xC021 to the PPP layer which processes the packet according to PPP LCP. For other supported protocol identifiers, the PPP layer removes the PPP encapsulation and passes the datagram and protocol identifier to the SNDCF. For unsupported protocol identifiers, the LCP protocol-Reject is passed to the RLP layer for transmission. The mobile station supports the PPP LCP Configure-Request, Configure-ACK, Configure-NAK, Configure-Reject, Terminate-Request, Terminate-ACK, Code-Reject, and Protocol-Reject. Other LCP packet types may also be supported. The PPP LCP negotiates the following configuration options:

    ✗ **Async control character map.** The mobile station does not require any mapping of control characters. The base station may negotiate mapping of control characters.

    ✗ **Protocol field compression.** This option applies when the protocol number is less than 0xFF.

    ✗ **Address and control field compression.** This option applies when the protocol number is not 0xC021.

  The mobile station may support other configuration options (e.g., maximum receive unit, authentication protocol, link quality protocol, or magic number). When an option that is not supported is received, the Configure-Reject is sent as an indication to the peer.

- **The Radio Link Protocol Layer** provides an octet stream service over the forward and reverse traffic channels and substantially reduces the error rate typically exhibited by these channels. This service is used to carry the variable-length data packets on the PPP layer. The RLP divides the PPP packets into TIA IS-95A traffic

channel frames for transmission. There is no direct relationship
between PPP packets and traffic channel frames. A large packet
may span multiple traffic channel frames, or a single traffic channel
frame may contain all or part of several small PPP packets. The
RLP is unaware of higher-layer framing; it operates on a featureless
octet stream, delivering the octets in the order received from the
PPP layer. For service options supporting an interface with multiplex option 1, RLP frames may be transported as primary or secondary traffic or as signaling via data burst messages. For the
primary or secondary traffic, the RLP generates and supplies
exactly one frame to the multiplex sublayer every 20 ms. The frame
contains the service option information bits. The multiplex sublayer
in the mobile station categorizes every received traffic frame and
supplies the frame type and accompanying bits, if any, to the RLP
layer. The frame type and frame category for primary and secondary traffic are given in tables 10.4 and 10.5. A blank frame is used
for blank and burst transmission of signaling traffic.

The signaling subchannel may carry frames from multiple RLPs,
with each RLP having a distinct BURST_TYPE. Each service

Table 10.4 RLP Frame with Primary Traffic

RLP Frame Type	Bits/Frame	Multiplex Option 1 Frame Categories
Full rate	171	1
Half rate	80	2, 6, 11
1/8 rate	16	4, 8, 13
Blank	0	5, 14
Erasure	0	All Others

Table 10.5 RLP Frame with Secondary Traffic

RLP Frame Type	Bits/Frame	Multiplex Option 1 Frame Categories
Rate 1	168	14
Rate 7/8	152	13
Rate 3/4	128	12
Rate 1/2	88	11
Blank	0	1–8
Erasure	0	9, 10

option defines a unique BURST_TYPE used for RLP. Each primary and secondary multiplex subchannel supports at most a single RLP layer. RLP data frames sent on one multiplex subchannel are not to be transmitted on another subchannel. RLP frames are not sent on the access and paging channels.

- **The Radio Interface** provides the physical layer between the mobile station and base station. In addition, it provides the multiplex sublayer, the radio link management, and the call control as defined in TIA IS-95A. The mobile station and the base station use service option 4 for async data services and service option 5 for Group-3 FAX services (see table 3.3). The mobile station and the base station do not transmit 1/4 rate frames when service option 4 or 5 is active. Service options 4 and 5 support an interface with multiplex option 1. RLP frames for service options 4 and 5 are transported only as primary traffic or signaling traffic. The mobile station and the base station perform service option negotiation for service options 4 and 5 as described in TIA IS-95A (sections 6.6.4.1.2 and 7.6.4.1.2). Initialization and connection in the mobile station and the base station to accept service option 4 or 5 in response to a Service Option Request Order are performed according to the specifications in TIA IS-99 (sections 3.8.4.1 and 3.8.4.2).

10.6.2 Short Message Service

The short message service [16] allows the exchange of short alphanumeric messages between a mobile station and the cellular/PCS system and between the cellular/PCS system and an external device capable of transmitting and optionally receiving short messages. The external device may be a voice telephone, a data terminal, or a short message entry system. The SMS consists of message entry features, administration features, and message transmission capabilities. These features are distributed between a cellular/PCS system and the SMS message center (MC), which together make up the SMS system. The MC may be either separate from or physically integrated into the cellular/PCS system.

Short message entry features are provided through interfaces to the MC and the mobile station. Senders use these interfaces to enter short messages, intended destination addresses, and various delivery options. MC interfaces may include features such as audio response prompts and DTMF reception for dial-in access from voice telephones, as well as appropriate menus and message entry protocols for dial-in or dedicated data terminal access. Mobile station interfaces may include keyboard and display features to support message entry. Also, a cellular voice service subscriber can use normal voice or data features of the mobile station to call an SMS system to enter a message.

An SMS teleservice can provide the option of specifying priority level, future delivery time, message expiration interval, or one or more of a series of short, predefined messages. If supported by the teleservice, the sender can request acknowledgment that the message was received by the mobile station. An SMS recipient, after receiving a short message, can manually acknowledge the message. Optionally, the recipient can specify one of a number of predefined messages to be returned with acknowledgment to the sender.

SMS administration features include message storage, profile editing, verification of receipt, and status inquiry capabilities. The SMS transmission capabilities provide for the transmission of short messages to or from an intended mobile station and the return of acknowledgments and error messages. These messages and acknowledgments are transmitted to or from the mobile station whether it is idle or engaged in a voice or data call. The cellular service provider may offer SMS transmission to its cellular voice and data customers only or may provide an SMS only service without additional voice or data transmission capability. All available mobile stations on a CDMA paging channel can receive a broadcast message. A broadcast message is not acknowledged by the mobile station. Broadcast messaging services may be made available to mobile stations on a CDMA paging channels as well as to mobile stations during a call on a CDMA traffic channel.

Figure 10.5 shows the network reference model for SMS. The base station contains the transceiver equipment, mobile switching center, and any interworking function (IWF) required for network connection. These elements are grouped together because there is no need to distinguish them. The MC element in the model represents a generic SMS message center function. The N reference point represents one or more standardized interfaces between an SMS message center and a BS. The terminal equipment (TE) is voice or data equipment connected either directly or indirectly to the MC. It is possible for the MC to be included in or colocated with a BS. In this case, the N interface is internal to the BS.

The SMS protocol stack for the CDMA mode of operation is shown in figure 10.6. The SMS bearer service is the portion of the SMS system responsible for delivery of messages between the MC and mobile user equipment. The bearer service is provided by the SMS transport layer and SMS relay layer.

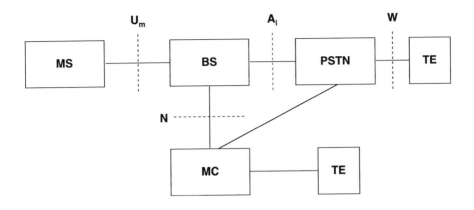

MS: Mobile Station

BS: Base Station

MC: Message Center

PSTN: Public Switching Telephone Network

TE: Terminal Equipment

Figure 10.5 Simplified SMS reference model. (Reproduced under written permission of the copyright holder [TIA].)

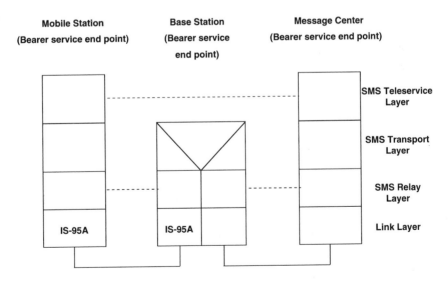

Figure 10.6 SMS protocol stack. (Reproduced under written permission of the copyright holder [TIA].)

The SMS transport layer is the highest layer of the bearer service protocol. The transport layer manages the end-to-end delivery of messages. In an entity serving as a relay point, the transport layer is responsible for receiving SMS transport layer messages from an underlying SMS relay layer, interpreting the destination address and other routing information, and forwarding the message via an underlying SMS relay layer. In entities serving as end points, the transport layer provides the interface between the SMS bearer service and SMS teleservice.

SMS uses the following layers:

• **The SMS Relay Layer** provides the interface between the transport layer and the link layer used to carry short message traffic. On the U_m interface, the SMS relay layer supports the SMS transport layer by providing the interface to the TIA IS-95A transmission protocols required to carry SMS data between CDMA mobile stations and the base stations. On the N interface, the SMS relay layer supports the SMS transport layer by providing the interface to the network protocols required to carry SMS data between the MC and TIA IS-95A base stations. The N reference point is assumed to be an intersystem network link with connectivity to the MC. Intersystem links can use a variety of public and private protocols. SMS protocols and message formats on intersystem links may differ from those used on the CDMA air interface. The N interface relay layer is responsible for formatting and parsing SMS messages as necessary when transmitting and receiving messages on the intersystem links. The SMS relay layer performs the following functions:

✗ Accepting transport layer messages and delivering them to the next indicated relay point or end point.

✗ Providing error indications to the transport layer when messages cannot be delivered to the next relay point or end point.

✗ Receiving messages and forwarding them to the transport layer.

✗ Interfacing to and controlling the link layer used for message relay.

✗ Formatting messages according to the SMS standards and/or other message standards, as required by the link layer and/or peer SMS layer.

• **The SMS Transport Layer** resides in SMS bearer service end points and relay points. In a bearer service end point, the SMS transport layer provides the means of access to the SMS system for teleservices that generate or receive SMS messages. In a bearer service relay point, the transport layer provides an interface between relay layers. The SMS transport layer uses relay layer services to originate, forward, and terminate SMS messages sent between mobile stations and MCs. It is assumed that the link layers used by the relay layers support message addressing, so that certain address parameters can be inferred by the relay layer from link layer headers and are therefore not required in transport layer messages. It is assumed that an SMS point-to-point message does not require certain address parameters because the link layers will provide this address. On the CDMA paging channel, for example, it can be assumed that the relay layer can extract the address from the ADDRESS field of the TIA IS-95A data burst message. SMS transport layers have different functions in SMS bearer service end points and relay points. In an SMS bearer service end point, the transport layer provides the following functions:

✗ Receiving message parameters from SMS teleservices, formatting SMS transport layer messages, and passing the message to the relay layer using the appropriate relay layer service primitives.

✗ Informing the relay layer when all expected acknowledgments of submitted messages have been received.

✗ Informing the teleservices when relay layer errors are reported.

✗ Receiving SMS messages from the relay layer and passing the messages to the SMS teleservice.

✗ Performing authentication calculations in mobile stations.

In an SMS bearer service relay point, the transport layer provides the following functions:

✗ Receiving SMS messages from a relay layer, reformatting the SMS transport layer message if necessary, and passing the message to another relay layer using the appropriate relay layer service primitives.

✗ Passing confirmations or error reports between the relay layers if requested.

✗ Performing authentication calculations or interfacing to the entities performing authentication calculations in the TIA IS-95A base stations.

The transport layer requires the following services from the relay layer:

✗ Accepting transport layer messages and delivering them to the next indicated relay point or end point.

✗ Returning confirmations or error reports for messages sent.

✗ Receiving messages and forwarding them to the transport layer with the appropriate parameters.

• **The SMS Teleservice Layer** resides in a bearer service end point and supports basic SMS functions through a standard set of subparameters of the transport layer's bearer data parameter.

When a mobile station sends an SMS User Acknowledgment message, the teleservice layer performs the following:

✗ Supplies the Destination Address parameter to the transport layer and sets the Destination Address parameter equal to the address contained in the Originating Address field of the SMS message being acknowledged.

✗ Sets the MESSAGE_ID field of the Message Identifier subparameters to the value of MESSAGE_ID field in the SMS message being acknowledged.

Broadcast Messaging Service Teleservice messages are sent using the SMS Deliver message. For more details, refer to TIA IS-637.

10.6.3 Packet Data Services for CDMA Cellular/Personal Communications Systems [17]

The packet data service option is specified by a mobile station at the time a packet link layer connection is opened. The packet data service option to be used may be configurable or may be fixed by the mobile station manufacturer. Service option 7 (see table 3.3) is used to request packet data service through an interworking function supporting an Internet standard PPP interface to network protocols. Service option 8 is used to request packet data service through an IWF supporting CDPD data services over a PPP interface. Even though service option 8 for packet data service does not use the same air interface as CDPD, this service options uses the same IWF supporting CDPD data services over PPP interface. Additional packet data service options may be defined in the future for the purpose of selecting other types of IWF resources or services requested.

Packet data service options provide a means of establishing and maintaining traffic channels for packet data service. When no other service option is connected, packet data service is carried as primary traffic. When another compatible service option is connected (e.g., voice or asynchronous data), packet data service can be carried as either primary or secondary traffic.

Figure 10.7 shows the reference model for packet data services. This reference model does not address intersystem and mobility issues. The reference points follow:

- **Reference Point R_m**—a physical interface to connect type 2 terminal equipment (TE2) to a mobile station (MT2).
- **Reference Point U_m**—a physical interface to connect mobile stations (type MT0 or MT2) to a base station and mobile switching center. This is the air interface.
- **Reference Point L**—A physical interface connecting a mobile switching center to an IWF.
- **Reference Point P_i**—a physical interface connecting an IWF to the public packet data network (PPDN).

The relay layers provide lower-layer communication and packet framing between the entities of the packet data service reference model. Over the R_m interface between TE2 and the MT2, the relay layer is a RS-232 [15] interface. Over the U_m interface, the relay layer is a combination of RLP and TIA IS-95A protocols. On the L interface, the relay layer uses

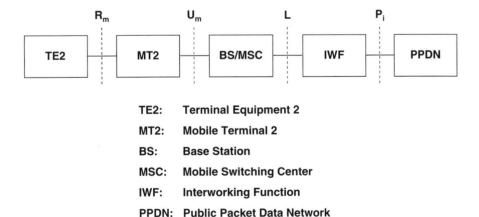

Figure 10.7 Reference model for packet data service.

the protocols defined in TIA IS-687 [18]. The two options for packet protocol stacks are shown in figures 10.8 and 10.9.

10.6.3.1 Relay Layer R_m Interface Protocol Option   The relay
layer R_m interface protocol option supports terminal equipment (type
TE2) applications in which the end-to-end link layer supports network

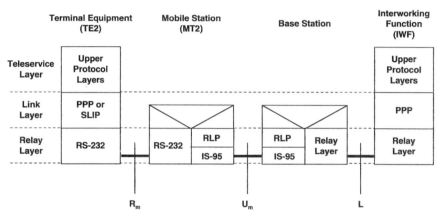

PPP: Point-to-Point Protocol
RLP: Radio Link Protocol

Figure 10.8 Relay layer R_m interface protocol option. (Reproduced under written permission of the copyright holder [TIA].)

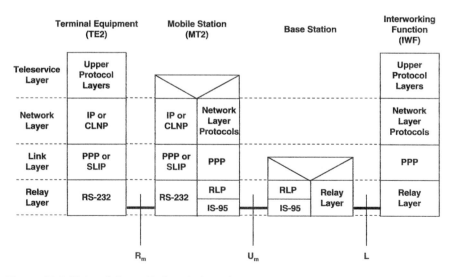

Figure 10.9 Network layer R_m interface option. (Reproduced under written permission of the copyright holder [TIA].)

and higher-layer protocols that reside entirely within the terminal equipment. With this option, the terminal equipment is responsible for all aspects of packet data service mobility management and network address management. For this R_m interface protocol option, the link layer provides the means for carrying packet data service between the TE2 and the IWF. For service options 7 and 8, the link layer is implemented using the Internet PPP. The link layer also provides protocol discrimination and supports functions such as header compression for network layer protocols.

For the R_m interface protocol option, the packet teleservice layer includes the network and higher-layer protocols used for communication between peer application entities in the TE2 and remote terminal equipment via the PPDN. For service option 7, network layer protocols may include all those protocols for which PPP indentifiers have been assigned by the Internet Assigned Number Authority. For service option 8, network layer protocols may include those protocols supported by the CDPD network.

10.6.3.2 Network Layer R_m Interface Protocol Option   The network layer R_m interface protocol option supports terminal equipment (type TE2) applications, such as CDPD applications, in which the mobile station (type MT2) supports network and higher-layer protocols. With this protocol option, the mobile station is responsible for all aspects of packet mobility management and network address management, and the terminal equipment application operates as if it were locally connected to a network layer routing server. In this protocol option, independent link layers provide the means to carry packet data service data between the terminal and the mobile station and between the mobile station and the IWF. For service options 7 and 8, the link layer between the mobile station and the IWF is implemented using the Internet PPP. In the MT2-IWF link layer, the link layer also provides protocol discrimination and supports functions such as header compression for network layer protocols.

The link layer between the MT2 and TE2 is implemented using the Internet PPP. Alternatively, the SLIP protocol, as defined in RFC 1055, may be used between the MS and the terminal to support the IP network layer protocol. For this R_m interface protocol option, the network layer also provides independent services between the terminal and the MS and between the MS and the IWF. The network layer between MS and IWF includes routing protocols (e.g., IP and CLNP) and the CDPD mobile network registration protocol (MNRP). The network layer between terminal and mobile station includes the routing protocols only.

With this R_m interface protocol option, the packet teleservice layer includes the transport and higher-layer protocols used for communication between peer application layer entities in the TE2 and remote terminal equipment via PPDN. Transport layer protocols may include protocols such as TCP or TP4.

10.6.3.3 Packet Data Service States   Packet data service has two states, active and inactive. Packet data service becomes active whenever selected by the user. The means for activating packet data service can be through commands on the R_m interface, a separate control on the user interface of the mobile station, or other means. If no user-accessible means of control is provided, packet data service is always active. When packet data service is active, the mobile station maintains the packet data service enabled status. The packet data service enabled status is used to control repeated packet data service origination attempts when packet data service is not available. When packet data service is enabled, the mobile station attempts to originate a packet data service with a serving IWF. When packet data service is disabled, the mobile station inhibits repeated attempts to originate packet data service.

10.6.3.4 Link Layer   The IWF and the mobile station maintain a link layer connection for the purpose of transmitting and receiving packet data between the mobile station and the IWF. The link layer connection is opened when a call is made using a packet data service option. Once a link layer connection is opened, bandwidth (in the form of a traffic channel assignment) is allocated to the connection on an as-needed basis. After opening a link layer connection, the mobile station and the IWF perform link layer configuration negotiation. The mobile station and IWF may also negotiate network layer protocol control parameters and perform procedures associated with network registration and authentication.

10.6.3.5 Link Layer Connection State   The link layer connection can be in any of the following states:

- **Closed.** The link layer connection is closed when IWF has no link layer connection state information about the mobile station.
- **Opened.** The link layer connection is opened when the IWF has link layer connection state information for the mobile station. The opened state has two substates:

✗ *Active.* An opened link layer connection is active when IWF has link layer connection state information for the mobile station, there is an *L* interface frame relay-switched virtual circuit for the mobile station, and the mobile station is on a traffic channel supporting a packet data service option.

✗ *Dormant.* An opened link layer connection is dormant when there is no *L* interface frame relay-switched virtual connection for the mobile station, and the mobile station is not on a traffic channel supporting a packet data service option.

The mobile station may not be aware of the state of the link layer connection. When the relay layer R_m interface protocol option is selected, the MT2 is never aware of the link layer connection state. When the network layer R_m interface protocol option is selected, the MS is aware of the link layer connection state while it is opened and active, but it may not be aware of a possible IWF-initiated link layer connection closure while it is dormant.

The mobile station and BS/MSC send packet data on the traffic channel using the RLP. RLP can be carried either as primary or secondary traffic. The mobile station and the BS/MSC support the physical layer, multiplex options, radio management, and call control protocols. When a packet data service option is connected as primary traffic, and no service option is connected for secondary traffic all secondary traffic data are discarded. For more details, refer to TIA IS-657 [17].

10.7 SUMMARY

Because wireless data networks do not operate without interconnection to other networks, in this chapter, we presented a variety of wireless data systems including the wide area wireless data systems and high-speed WLANs and the specific systems supported by CDMA. We examined the various standards being adopted by the IEEE, the WINForum, HIPERLAN (in Europe), and ARPA for wireless LANs. Since packet networks are an important part of wireless networks, we briefly stated the characteristics of the access methods in common use and defined their throughput equations. The common packet protocols are ALOHA, slotted-ALOHA, and CSMA/CD.

We then presented the methods used to control errors for wireless data systems. We concluded by presenting the highlights of the TIA IS-99, TIA IS-637, and TIA IS-657 standards for CDMA cellular systems. CDMA supports asynchronous data, facsimile, packet data, and short message service to end points in another wireless network or to the wireline network. We examined the reference models and protocol stacks for each of these data services.

10.8 REFERENCES

1. Bates, R. J., *Wireless Networked Communication*, McGraw-Hill, New York, 1994.
2. Hammond, J. L., and O'Reilly, J. P., *Performance Analysis of Local Computer Networks*, Addison-Wesley Publishing Company, Reading, MA, 1986.
3. Habab, I. M., Kavehrad, M.,and Sundberg, C.-E. W., "ALOHA with Capture Over Slow and Fast Fading Radio Channels with Coding and Diversity," *IEEE Journal of Selected Areas of Communications*, 6, 1988, pp. 79–88.
4. Pahlavan, K., and Levesque, A. H., *Wireless Information Networks*, John Wiley and Sons, New York, 1995.
5. RFC 791, "Internet Protocol."
6. RFC 793, "Transmission Control Protocol."
7. RFC 792, "Internet Control Message Protocol."
8. RFC 1122, "Requirements for Internet Hosts—Communication Layers."
9. RFC 1144, "Compressing TCP/IP Headers for Low-Speed Serial Links."
10. RFC 1332, "The PPP Internet Protocol Control Protocol (IPCP)."
11. RFC 1661, "The Point-to-Point Protocol (PPP)."
12. RFC 1700, "Assigned Numbers."
13. TIA IS-95A, "Mobile Station—Base Station Compatibility Standard for Dual-Mode Wideband Spread Spectrum Cellular System."
14. TIA IS-99, "Data Services Option Standard for Wideband Spread Spectrum Digital Cellular System."
15. TIA IS-232-E, "Interface Between DTE and DCE Employing Serial Binary Data Interchange."
16. TIA IS-637, "Short Message Services for Wideband Spread Spectrum Cellular Systems."
17. TIA IS-657, "Packet Data Services Option for Wideband Spread Spectrum Cellular System."
18. TIA IS-687, "Data Services Inter-Working Function Interface Standard for Wideband Spread Spectrum Digital Cellular System."
19. Skalar, B., *Digital Communications–Fundamentals and Applications*, Prentice Hall, Englewood Cliffs, NJ, 1988.
20. Viterbi, A. J., and Padovani, R., "Implications of Mobile Cellular CDMA," *IEEE Communication Magazines*, 30, (12), pp. 38–41, 1992.
21. Walrand, J., *Communications Networks: A First Course*, Irwin, Homewood, IL, 1991.
22. Woo, I. and Cho, D.-H., "A Study on the Performance Improvements of Error Control Schemes in Digital Cellular DS/CDMA Systems," *IEICE Trans. Communications*, E77-B, (7), July 1994.

Management of CDMA Networks

11.1 INTRODUCTION

In the preceding chapters, we discussed the elements of a cellular and PCS system with CDMA technology to provide service to a subscriber. In this chapter, we present those elements of the CDMA system that keep the system operating on a day-to-day basis. These elements are referred to as the operations, administration, maintenance, and provisioning systems. With OAM&P systems, cellular and PCS service providers monitor the health of all network elements, add and remove equipment, test software and hardware, diagnose problems, and bill subscribers for services.

In the past, management of telecommunications networks was simple. In the days prior to the deregulation and privatization of the telephone industry, the service provider dealt with fewer issues. Generally, the competitive pressure was less than it has been after deregulation. The networks were composed of equipment from fewer vendors, thus there were fewer multivendor management issues. As service providers moved into a mixed vendors' environment, they could no longer afford different systems for each network element. The standards groups in committee T1M1 and ITU defined the interface and protocols for operations support systems under the umbrella of the Telecommunications Management Network (TMN).

Service providers are deploying PCS into the existing cellular competitive environment. Therefore, they must offer cost-effective services. Even though management functions are necessary for the smooth operation of a system, they are a cost of doing business and do not directly improve the bottom line of a business. Therefore, service providers must manage the cost of these systems. Considerations of initial cost of the management systems and their operational costs must be carefully examined. A key for the success of personal communications systems lies in the system operator's ability to manage the new equipment and services effectively in a mixed vendors' environment. For PCS to succeed, the OAM&P costs for PCS networks and services must be lower than those of either the existing wireless or wireline networks. Automation is required to develop machine-to-machine interfaces to replace many manual functions. The need to manage mixed vendors' equipment requires some form of standardization. Finally, the need to support rapid technological evolutions demands that the interfaces be general and flexible. Furthermore, to ensure that the interfaces have sufficient consistency to allow some level of integrated management, it is necessary to develop a set of guiding principles. That set of guiding principles is provided by TMN.

11.2 MANAGEMENT GOALS FOR PCS NETWORKS

The original cellular systems were purchased from equipment vendors as complete and proprietary systems. The interfaces between network elements were proprietary to each vendor, and network elements from one vendor did not work with network elements from another vendor. Cellular systems included management systems that were tailored to the vendor's hardware and software. Both cellular and PCS networks are migrating to a mixed vendor environment with open interfaces. Thus, a service provider can purchase switching equipment from one vendor, radio equipment from another vendor, and network management equipment from a third vendor. In this new environment, specific goals for network management hardware and software are necessary. These network management goals for PCS follow.

Operation in a Mixed Vendor Environment. The network management hardware, software, and data communications must support network elements from different vendors. It should allow for the integration of equipment from different manufacturers in the same TMN, by using clearly defined interworking protocols.

Availability of Multiple Solutions. Service providers should be able to purchase network management solutions from multiple and competitive vendors to ensure the lowest cost and richest feature set for the system.

Use of Existing Resources. Cellular companies and existing wireline companies should be able to manage their PCS network with minimal additions of new data communications, computing platforms, and so on. If possible, minimal additions should be made to existing distribution networks to implement the required control or reporting system.

Support for Multiple and Interconnected Systems. When a service provider has deployed multiple systems, the different management systems should interact effectively and efficiently to deliver service by sharing a common view of the network.

Support for Sharing System and Information Among Multiple Service Providers. Multiple service providers must interact to share information on billing records, security data, subscriber profiles, and so forth, and to share call processing (e.g., intersystem handoffs). The network management systems, therefore, should allow flexible telecommunication management relationships among multiple service and network providers of PCS. Flexible management relationships include complex network and service provider arrangements consistent with individual providers that operate as separate business entities. Wireless access providers might also want management access through intermediate networks.

Support for Common Solutions Between End Users and Service Providers. Many services require a joint relationship between the service provider and the end user. End-to-end data communication is the most common example of the need for a joint relationship. When service providers and end users are operating in this mode, their management functions should be interconnected. The solutions should allow flexible telecommunication management relationships between service providers and end users.

Transparent. A wireless network management system should be as transparent as possible to the technology used in wireless network implementation.

Flexible. The network management system should be flexible to allow for evolution in wireless network functions and services.

Modular. The network management system should be modular so that irrespective of the future network size or the location of control and knowledge, the management functionality can support all management aspects.

Fail-Safe. Neither equipment failure nor operator error should render the management system and/or the wireless network inoperative.

11.3 REQUIREMENTS FOR MANAGEMENT OF PCNs

A personal communication network is a telecommunications network where users and the terminals are mobile rather than fixed. In addition, there are radio resources that must be managed and do not exist in a wireline network. Furthermore, the network model can be a cellular model where the radio and switching resources are owned and managed by one company. A second newer model has the wireline company operating the standard switching functions and a wireless company operating the radio-specific functions. With this new mode of operation, network management requirements for personal communication networks include standard wireline requirements and new requirements specific to personal communication networks. The new personal communication network-specific requirements are needed for the following areas.

Management of Radio Resources. PCS allows terminals to connect to the network via radio links. These links must be managed independent of the ownership of the access network and switching network. The two networks may consist of multiple network service providers or one common service provider. The multiple operator environment will require interoperable management interfaces.

Personal Mobility Management. Another important aspect of personal communication networks is mobility management. Users are no longer in a fixed location but may be anywhere in the world. Users can be addressed by their identifications (IDs) without needing to take into account their current location or status. The network automatically finds them and correctly routes the call. The network also recognizes the originator's ID and delivers stored information that augments the user's identity. This will increase the load on the management system as users manage various decision parameters about their mobility. For example, they may request different services based on the time of day or terminal busy conditions.

Terminal Mobility Management. The primary focus of personal communication networks is to deliver service via wireless terminals. These terminals may appear anywhere in the worldwide wireless network. Single or multiple users may register on a wireless terminal. The terminal associated with a user's ID is tracked, perhaps across networks, in order to deliver messages or calls to the personal ID. Terminal mobility is the ability of a terminal to access telecommunications services from different locations while in motion. The terminal management function may be integrated with the personal management function or may be separate from the personal management function. Different service providers may operate their system in a variety of modes. The management function must support all modes.

Service Mobility Management. Service mobility is the ability to use today's vertical features from remote locations or while in motion. As an example, the user has access to the messaging service anywhere, anytime. The user also has access to a global locating service. The ability to specify an event is provided via a user interface that is flexible enough to support a number of input formats and media. A user can specify addressing in a simple, consistent way no matter where he or she is.

11.4 TELECOMMUNICATION MANAGEMENT NETWORK AND WIRELESS NETWORK MANAGEMENT

TMN includes a logical structure, which originates from data communication networks, to provide an organized architecture for achieving the interconnection between various types of operations systems and/or telecommunication equipment types. Management information is exchanged between operations systems and equipment using an agreed upon architecture and standardized interfaces, which include protocols and messages. The Open System Interconnect management technology has been chosen as the basis for the TMN interface. The principles of TMN provide for management through the definition of a management information model (managed objects), which is operated over standardized interfaces. This model is developed through the definition of a required set of management services that are then decomposed into various service components and then decomposed into management functions. An information model represents the system and supports the management functions.

Three important aspects of the TMN that can be applied to manage wireless networks follow:

- A layered architecture that has five layers: business management layer (BML), service management layer (SML), network management layer (NML), element management layer (EML), and network element layer (NEL).
- A functional architecture that defines functional blocks [operations systems functions (OSF), mediation function (MF), work station function (WSF), network element function (NEF), and Q adapter function (QAF)] is shown in figure 11.1.

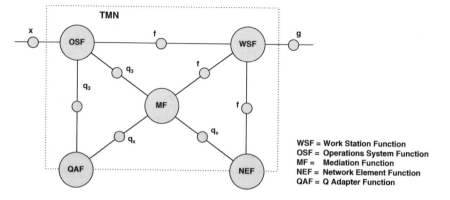

WSF = Work Station Function
OSF = Operations System Function
MF =   Mediation Function
NEF = Network Element Function
QAF = Q Adapter Function

Figure 11.1 Functional architecture of TMN.

• A physical architecture that defines management roles for operations systems, communication networks, and network elements is given in figure 11.2.

TMN functions can be divided into two classes: basic functions and enhanced functions. *Basic functions* are used as building blocks to implement enhanced functions such as service management, network restoration, customer control/reconfiguration, and bandwidth management. TMN basic functions are further grouped into three categories.

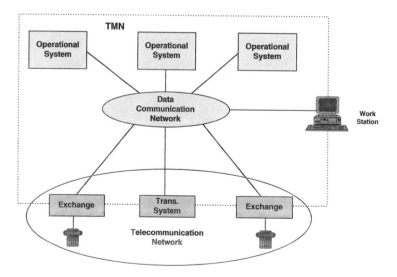

Figure 11.2 Physical architecture of TMN.

I. Management Functions

Accounting Management

Collect accounting (billing) data

Process and modify accounting records

Performance Management

Data collection

Data filtering

Trend analysis

Fault Management

Alarm surveillance (analyze, filter, correlate)

Fault localization

Testing

Security Management

Provide secure access to network element functions/capabilities

Provide secure access to TMN components (e.g., operations systems, SubNetwork Controllers (SNC), Mediation Device (MD), etc.)

Configuration Management

Provisioning

Status

Installation

Initialization

Inventory

Backup and restoration

II. Communication Functions

Operations System/Operations System Communications

Operations System/Network Element Communications

Network Element/Network Element Communications

Operations System/Workstation (WS) Communications

Network Element/WS Communications

III. Planning Functions

Network Planning, including physical resource (facility, equipment, etc.) planning, Workforce Planning.

11.4.1 Functional Architecture

The TMN architecture provides a degree of flexibility to support various and varying network topologies. The TMN functional and reference models define the key components of a management solution. The TMN models define a number of functional components that can be used by TMN designer and can be mapped onto physical TMN elements. The components of the functional architecture follow (refer to fig. 11.1).

Operations Systems Function. The OSF processes information related to telecommunications management for the purpose of monitoring, coordinating, and/or controlling telecommunication functions. The objects that are under the control of a given OSF are the components of the OSF management "domain." In certain cases, the management information may be partitioned into layers that can be hierarchically organized. In such cases, different OSFs may be responsible for the different layers that represent their respective management domains. This type of implementation is referred to as a "logically" layered architecture.

The physical implementation of OSFs must provide the alternatives of either centralizing or distributing the general functions including:

- Support application programs,
- Database functions,
- User terminal support,
- Analysis programs,
- Data formatting and reporting, and
- Analysis and decision support.

Reference Points. The relationships between components of the TMN functions are defined by reference points that allow identification of the boundaries between wholly self-contained functional units. The TMN standard interfaces follow:

- q_3 **Interface** supports the full set of OAM&P functions between operations systems and network elements and between operations systems and mediation devices. The q_3 interface may also be capable of supporting the sets of OAM&P functions between operations systems.
- q_x **Interface** supports a full set or a subset of OAM&P functions between mediation devices and network elements, between mediation devices, and between a network element with mediation function and another network element.

- **x Interface** supports the set of operations systems to operations systems functions between TMN networks or between a TMN network and the x interface on any other type of management network.
- **f Interface** supports the set of functions for connecting workstations to operations systems, mediation devices, or network elements through a data communication network.
- **g Interface** is located outside of TMN and is between users and the workstation function.
- **m Interface** is located outside TMN and is between Q adapter functions and non-TMN managed entities.

Data Communication Function. The DCF is implemented via a data communication network (DCN). The DCN helps to connect the various TMN components with one another. The DCN is used when various functional groupings are implemented remotely from others. Each functional component contains a message communication function (MCF) to allow connection to the data communication function provided by the DCN.

Mediation Function. The MF primarily routes and/or acts on information passing between standardized interfaces. MFs can be located at network element(s) and/or operations system(s). The processes that can form mediation are classified into the following process categories:

- Communication control,
- Protocol and data conversion,
- Communication (passing) of primitive functions,
- Processes involving decision making, and
- Data storage.

Network Element Functions. NEFs communicate with the TMN for the purpose of being monitored and/or controlled. The NEF provides the telecommunications and support functions that are required by the telecommunications network being managed. The NEF includes the telecommunications functions that are the subject of management. These functions are not part of TMN but are represented to the TMN by the NEF. The part of the NEF that provides this representation in support of the TMN belongs to the TMN itself, whereas the telecommunication functions themselves are external to the TMN.

Q adapter Function. The QAF is used to connect those non-TMN entities, which are NEF-like and OSF-like to the TMN architecture. The job of the QAF is to translate messages between a TMN reference point and a non-TMN (e.g., proprietary) reference point.

Workstation Functions. WSFs are defined as the functionality that provides interaction between craft personnel and the OSFs.

11.4.2 Physical Architecture

An example of a simplified architecture for a TMN is shown in figure 11.2. TMN functions can be implemented in a variety of physical configurations. The relationship of functional blocks to TMN building blocks is given in table 11.1.

Table 11.1 Relationship of the Functional Blocks to TMN Building Blocks

Device	NEF	MF	QAF	OSF	WSF
Network Element	M	O	O	O	O
MD	—	M	O	O	O
QA	—	—	M	—	—
Operations System	—	O	O	M	O
WS	—	—	—	—	M

M = Mandatory; O = Optional

11.4.2.1 Layered Architecture

TMN uses a layered architecture; the major functionality of the five layers follow (refer to fig. 11.3).

Business Management Layer. The BML has the responsibility for the total enterprise and supports agreements between operators. This layer normally carries out goal-setting tasks rather than goal achievement but can become the focal point for action in cases where executive action is necessary. The BML is part of the overall management of the enterprise, and many interactions with other management systems are necessary.

Service Management Layer. The SML is concerned with, and responsible for, the contractual aspects of services that are being provided to customers or available to potential customers. The five principle roles of the SML are:

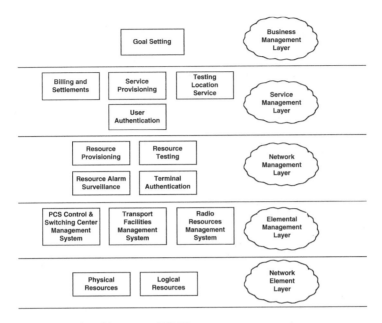

Figure 11.3 Layered architecture of TMN.

- Interfacing with customers and other administrations,
- Interacting with service providers,
- Interacting with the network management layer,
- Maintaining statistical data (e.g., quality of service), and
- Allowing interactions between services.

Network Management Layer. The NML is responsible for the management of all the network elements, as presented by the element management layer, both individually and/or as a set. This layer is not concerned with how a particular element provides service internally. Functions addressing the management of a wide geographic area are located at this layer. The whole network is completely visible, and a vendor independent view is maintained. The three principle roles of the NML are: (1) the control and coordination of the network view of all network elements within its domain, (2) the provision, cessation, or modification of network capabilities for the support of services to customers, and (3) interaction with the SML on performance, usage, availability, and so on. The NML provides the functionality to manage a network by coordinating all activities across the network and supporting the network demands made by the SML.

Element Management Layer. The EML manages each network element on an individual basis and supports abstraction of the functions provided by the network element layer. The EML has a set of element managers that are individually responsible for some subset of network elements. Each element manager has three principal roles:

- Control and coordinate of a subset of network elements,
- Provide a gateway (mediation) function to allow interactions among network elements, and
- Maintain a statistical log and other data about elements.

Network Element Layer. The NEL consists of logical and physical resources to be managed.

Information models to support the functions identified for various layers are developed in the O interface (refer to fig. 11.4) standards. These models will then be used to exchange management information across interfaces according to the particular physical architecture of the managed network. The information models for different levels of abstraction provide the required flexibility in developing implementation to support the NEL, EML, NML, SML, and BML functions.

The following five EML functions simplify management of a wireless network.

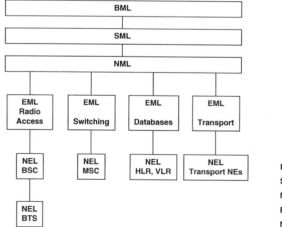

Figure 11.4 Layered architecture as applied to wireless network.

- **Aggregated Resource Provisioning.** This function allows NML (or other EML) applications to request certain resource provisioning functions on a collection of NEL resources within its span of control. This function performs these tasks on individual resources. This may result in simplification of the provisioning process.
- **Alarm Correlation and Filtering.** In many situations, a single resource failure may be the source of failures in multiple related resources. To reduce the burden on the network level alarm surveillance tasks, the notifications of these failures may be correlated and filtered, emitting only a single alarm from the EML.
- **Software Management.** EML software management functions enable the NML applications to manage wireless software in a more generic way, while the EML provides software management for wireless network elements within its span of control.
- **Auto Resource Discovery.** The EML should be able to detect the presence of newly added resources, such as the addition of a BTS to a BSC or changes in equipment (e.g., memory and power supply).
- **Network Resource Inventory Management.** This function provides management of the resource inventory for the wireless network elements within its span of control.

11.4.3 Quality of Service

The objective of a wireless network management system is to integrate the wide range of wireless network operator activities in order to achieve coherent and seamless information exchange at a given quality of service (QoS) objective while achieving business objectives. The wireless network operator typically establishes a QoS criteria and objective in the context of the service levels to be provided to customers and with knowledge of the performance of the network infrastructure. It is necessary to compare these objectives and expectations with experience gained by monitoring the performances of services according to the level of customer complaints. The operator must also monitor the technical performance of the network in order to initiate improvements in QoS. Even though the technical performance of the network can be monitored, it does not necessarily reflect the service performance that the customer sees.

ITU-T recommendations I.350 and I.140 provide the general aspects of QoS and network performance in digital networks, including ISDN. These recommendations are applicable to a PCS network, as PCS use both digital networks and ISDN and may also provide users with digital data services in the future.

The ITU-T recommendations provide descriptions of QoS and network performance parameters. In the ITU-T Recommendation I.350, QoS aspects are restricted to the parameters that can be directly observed and measured at the point at which the service is accessed by the customer. Network performance is measured in terms of the parameters that are meaningful to the service provider. Network performance parameters are used for system design, configuration, operation, and maintenance.

11.5 ACCOUNTING MANAGEMENT

Accounting management refers to the usage information generation and processing function that is used to render bills to the customers. It includes the distributed function that measures usage of the network by subscribers or by the network itself (e.g., for audit purposes). It also manages call detail information generated during the associated call processing to produce formatted records containing this usage data. The formatted accounting management records include billing systems, operation systems for performance management and fraud detection, and subscribers' profiles. The primary objective of the accounting management data is to render a bill to the customer using the wireless service. The customer of services may be another network or the end user.

Accounting for resource utilization consists of three subprocesses (see fig. 11.5):

- **The Usage-Metering Process** is responsible for the creation of usage-metering records generated by the occurrence of accountable events in the system (i.e., calls originated or received and requests for supplementary services). In general, use of a service requiring multiple resources will generate several usage-metering records.
- **The Charging Process (or Rating Process)** is responsible for collecting the usage-metering records that belong to a particular service transaction and for combining them into service transaction records (i.e., call detail records). Also, pricing information is added to the service transaction records. The rating process is also responsible for logging the service transaction records.
- **The Billing Process** is responsible for collecting the service transaction records, for selecting the records that belong to a particular subscriber during a particular time-period, and for producing the bill.

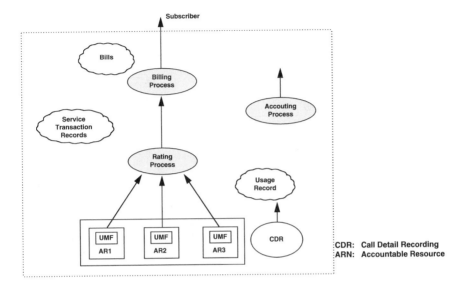

Figure 11.5 The accounting process.

11.5.1 Billing Data Management

Call and event data are required for a number of network management activities including, but not limited to, the following:

- Billing of home subscribers, either directly or via service providers, for network utilization charges;
- Settlement of accounts for traffic carried or service performed by fixed network and other operations;
- Settlement of accounts with other wireless networks for roaming traffic;
- Statistical analysis of service usage; and
- As historical evidence in dealing with customer service and billing complaints.

To support these activities, the accounting management system should support the following functions over the O interface:

- **Usage Management Functions**—usage generation, usage edit and validation of call events, usage error correction for call events, usage accumulation, in-call service request, service usage correlation, usage aggregation, usage deletion, usage distribution

- **Accounting Process Functions**—usage testing, usage surveillance, management of usage stream, administration of usage data collection
- **Control Functions**—tariff administration, tariff system change control, tariff class management, advice of charging management, data generation control, partial record generation control, data transfer control, data storage control, emergency call reporting

The PCS/cellular accounting management must also address the following items:

- **Distributed Collection of Usage Data.** Since many services will be "multinetwork services," it will be difficult for a single node (such as a switch) to generate a complete record of a call, as is done today. An example is roaming of mobile subscribers. This may involve multiple network nodes, possibly belonging to different service providers.
- **Improved Performance for Billing Collection and Report Generations.** Usage data are expected to be transmitted to a billing operations system in near-real time. There also will be concern about data concurrency and latency restrictions. These issues will pose significant performance requirements.
- **Multitude of Charging Strategies.** The usage data collected for billing must be flexible enough to support a variety of charging strategies. In some cases, the consumer will pay directly; in other cases the provider will pay. Charges for a call may be split between calling and called parties. In the case of multiple providers, the consumer may deal with only one provider (such as 900 services). Different legs of a call may receive different charging treatment.
- **Rapid Introduction of Diverse Services.** The accounting management structures must allow for the timely introduction of additional formats as services, technology, and pricing strategy evolution. When adding new formats, backward compatibility should be maintained.

11.5.2 Data Message Handling and TMN

Figure 11.6 shows the reference model for the data message handler (DMH) as defined in TIA IS-124, [9]. The DMH consists of the call detail information source (CDIS), call detail generation point (CDGP), call detail collection point (CDCP), and call detail rating point (CDRP). The functions of these components follow:

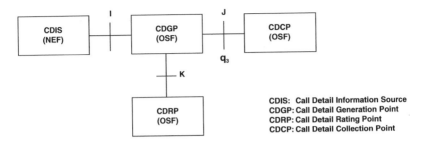

Figure 11.6 DMH reference model and mapping to TMN.

- **The CDIS** may be a PCS/cellular switching center, a radio port controller (RPC), a radio access system controller (RASC), a terminal mobility controller (TMC), a personal mobility controller (PMC), or any other source of call detail information. The CDIS provides call information without additional services or call detail information storage. In TMN, the CDIS function is associated with the NEF.

- **The CDGP** is responsible for collecting call detail information and encoding the messages for delivery to a CDCP. The CDGP may edit the collected call detail information as necessary to perform operations including reducing redundant information, discarding inconsistent records, eliminating data not required by the collection point, merging related call information, or adding rate information. The CDGP is also responsible for addressing the messages to the proper CDCP. Thus, it provides a distribution of usage data based on the service user. The CDGP stores call detail information until they are received and reconciled by the CDCP. These functions are OSF in TMN.

- **The CDCP** receives the call detail records. A CDCP may use or process call details. Examples of services are subscriber billing, intersystem charging, net settlement, and subscriber fraud detection. A CDCP may perform other record edits as necessary. The CDCP performs functions normally found in a billing system. In the context of the TMN, a CDCP performs OSF.

- **The CDRP** calculates and returns the currency information based on call detail information. In a public switched telephone network (PSTN) environment, this function is referred as tariffing. In the PSTN, the CDRP is frequently associated with the billing system and performs as an OSF.

The DMH is intended not only for subscriber billing but also to improve wireless services management such as fraud management, credit monitoring, performance management, and network planning. The DMH requires that real-time data on the elapsed air-time and tandem-time from all switches involved in a call be sent to the control point at call completion. The control point is the anchor system switch that makes the first radio contact with the subscriber directly or through other switches during the call. At least two call records are generated for a call delivered to a roaming subscriber. The home (or originating) system has the data about the call between the originating system and the system currently serving the subscriber. The visited or serving system directly records data about the subscriber's location, use of air-time, usage of features, etc. These records are correlated to generate a consolidated bill of a call event regardless of the number of cellular/personal communications systems involved in this call. The DMH specifies five types of call records: audit records, leg records, segment records, activity report records, and event report records. These records contain information about different aspects and portions of a call such that each can be cross-referenced to the other in order to generate the complete record for further inter administration processing. In figure 10.6, a mapping of the DMH reference points I, J, and K and processing entities CDIS, CDGP, CDRP, and CDCP to possible TMN reference points and functions is also given. At present, the I and K reference points defined in the DMH are proprietary interfaces. The data transferred across the J reference point are specified based upon an open interface.

The information collected by the DMH CDGP follows:

- **Chronology**

  ✗ *Time of Day*
  ✗ *Time Day Offset*—offset in minutes of local civil time with respect to universal coordinated time
  ✗ *Date*

- **Usage Measurement**

  ✗ *Duration*—measures the duration between a beginning event and an ending event in tenths of seconds.
- **Digits**

  ✗ *Account Code*

✗ *Billing Digits*—specify and identify the party to be billed under specified billing arrangements

✗ *Alternate Billing Digits*—identify the party to be billed under special billing arrangements

✗ *Called Digits*—specify and identify the called party

✗ *Calling Digits*—specify and identify the calling party

✗ *Destination Digits*—specify a telephone directory number toward which a call is routed over a network

✗ *Dialed Digits*

✗ *Interaction Digits*—specify digits dialed by the calling party in response to a voice or tone prompt

✗ *Routing Digits*—specify the digits used to provide a special routing for a call to access a trunk, steer a call, or specify a private network hop-off

- **Identifier**

  ✗ *Carrier Digits*

  ✗ *Billing Identification Number*—specify the number assigned at the anchor element, which was originally serving when it requested the call or was offered a call from the network

  ✗ *Electronic Serial Number*

  ✗ *Mobile Identification Number*

  ✗ *Related Billing Identification Number*

  ✗ *Report Identification Number*

  ✗ *Calling Party Indicator*

  ✗ *Calling Party Category*

  ✗ *Business Relation Identification*—specify a special billing relationship for purposes of cost allocation, revenue sharing, or similar purposes

- **Location Identifier**

  ✗ *H coordinate*

  ✗ *V coordinate*

  ✗ *Place Name*

  ✗ *North American Jurisdiction*

  ✗ *NPA-NXX*—6 digits to represent the rating point used by the serving point

  ✗ *Location Area Identifier*

  ✗ *Latitude*

  ✗ *Longitude*

✗ *Resolution*—specify the resolution of latitude and longitude measurements in feet

• **Network Resources**

✗ *Switch Number*
✗ *System Identification*
✗ *Feature Bridge Identification*
✗ *Trunk Group Identification*
✗ *Trunk Number*
✗ *Transceiver Number*
✗ *CDMA Station Classmark*
✗ *FDMA Station Classmark*
✗ *TDMA Station Classmark*

• **Features and Services**

✗ *Bearer Indicator*
✗ *Feature Bridge Indicator*
✗ *Feature Indicator*
✗ *Feature Operation*
✗ *Services Used Indicator*

• **Charging and Rating**

✗ *Billing Indicator*
✗ *Charge Amount*
✗ *Charge Indicator*
✗ *Charge Tax*
✗ *Charge Tax Indicator*
✗ *Charge Units*
✗ *Multiple Rating Period Indicator*
✗ *Rate Period Indicator*
✗ *Toll Tariff Indicator*

• **Event Indicator**

✗ *Authorization Type*
✗ *Authorization Count*
✗ *Event Indicator*
✗ *Feature Result Indicator*
✗ *Release Indicator*
✗ *Redirection Indicator*
✗ *Priority Access Indicator*

• **Traffic Measurement (see section 11.9.5)**

• **Fraud Indicator**

✗ *Credit Limit*

✗ *Call Count*

✗ *Number of Calls / Hour*

✗ *Call Patterns*

11.6 SECURITY MANAGEMENT

Security management contains a set of functions that control and administer the integrity, confidentiality, and continuity of telecommunication services against security threats. This set of functions support the application of security policies and audit trails by controlling security services and mechanisms, distributing security-related information, and reporting security-related events.

Security management does not address the passing of security-related information in protocols and services that call up specific security services (e.g., in parameters of connection requests).

11.6.1 Information-Gathering Mechanisms

It is desirable to record the occurrence of various security events. Depending on the type of information, frequency of occurrence, and importance of the event, one of several mechanisms may be used to record the occurrence:

• Scanners to collect and periodically report measurement information on high-frequency or low-importance events;

• Counters, allowing for the definition of threshold crossing and notification severity;

• Security alarms for high-importance and/or infrequent events.

11.6.2 Audit Trail Mechanism

Some security events that occur during the life of a system may need to be reviewed immediately, and immediate actions may need to be taken. For other events, it may be useful to review the history in order to identify patterns of failures or abuse. It is recommended that these data be maintained in a log that holds security audit records. This log may be kept either at the agent or at the manager side.

11.6.3 Security Alarm Reporting Mechanisms

The manager needs to be alerted whenever an event indicating a potential breach in security of the wireless network is detected. This detection may be reported by an alarm notification. The security alarm report should identify the cause of the security alarm, its perceived severity, and the event that caused it. All requirements relative to storage and forwarding of alarm notifications identified under fault management are assumed to apply to security alarms as well.

11.7 CONFIGURATION MANAGEMENT

Configuration management contains a set of functions that are used to control, identify, and collect data from/to the network elements. A wireless network operator needs to have an overview of the entire network including but not limited to the following:

* Hardware, firmware, software, and combinations that are compatible with each other.
* Hardware, firmware, and software that are currently used in each network element.
* Frequency plan for the wireless network and the frequency allocation for each cell.
* Coverage plan and the coverage area of each base station.

Configuration management is concerned with monitoring and control that relate to the operation of units in the system. It deals with initial installation, growth, or removal of system components. Configuration management supports the following management functions.

Creating Network Elements and Resources. The creation of a network element is used initially to set up a wireless network or to extend an already existing network. The action of creation includes a combination of installation, initialization, and introduction of the newly installed equipment to the network and to the operations system that will control it. The creation can affect equipment, software, and data. Whenever a wireless network or parts of it are installed, the created network elements are required to be:

* Physically installed and tested and then initialized with a possible default configuration;

- Logically installed by means of introduction to the network possibly involving changes to existing network element configuration;
- Put into service.

The management system must support mechanisms for user-friendly identification of these elements and resources. It should be possible to associate information such as resource name, location, description, and version with logical or physical elements.

The sequence of physical and logical installation may vary depending on the specific wireless network operator strategy. In case the logical creation occurs before the physical creation, no related alarms should be reported to the operator.

Deleting Network Elements and Resources. If a network is found to be overequipped, the operator may wish to reduce the equipment or to reuse the spare equipment elsewhere. This situation can occur when an operator overestimates the traffic in one area and underestimates the load in a different area. The deletion of a network element requires:

- Taking the affected network elements out of service;
- Logically removing it from the network (possibly involving changes to other network element configurations, for example, the neighbor cell description);
- If necessary, physical dismantling the equipment;
- Returning other affected network elements to service.

The sequence of logical and physical removal will not matter if the affected network elements are taken out of the service prior to their removal. This will help to protect the network from error situations.

Conditioning Network Elements and Network Resources. When a network element is to be modified, the following actions should be performed:

- Logical removal,
- Required modification, and
- Logical reinstallation.

This sequence is recommended to protect the network against fault situations that may occur during the modification process. The result of conditioning should be able to be determined by the operator by using the appropriate mechanisms.

A modification to data that has a controlling effect on some of the resources could affect the resource throughput or its capability to originate new traffic during the modification period. This should be evaluated for particular modifications because the capacity of the network element can be reduced without affecting the ongoing traffic. The forecast of a modification on capacity, throughput, and current activity of a resource will help the operator of a network to decide when the modification should be carried out.

The data characterizing a wireless network will not all be subject to the same rate of change or need to be modified using the same mechanism. Changes to the logical configuration may also need to be applied across multiple network elements.

A major aspect of configuration management is the ability of the operator to monitor the current configuration of the network. This is necessary to determine the operational state of the network and to determine the consistency of information among various network elements. The monitoring capability requires the information request function, the information report function, and the response/report control function.

- **Information Request Function.** The network operator should monitor the network in order to support its operation. The operations system should be able to collect information as needed from various network elements. The operations system should be able to request information for any single attribute defined in a management information base. Also, the operations system should be able to collect a large amount of data in a single request by providing an appropriate scope and filter in the request. On receipt of a valid request, the addressed network element must respond with the current values of the specified data elements. This response should be immediate if so requested by the operations system. However, in the cases where very large amounts of data are concerned and where the operations system and the network element support the capabilities, the operations system may request the network element to store the data in a file and transfer it using a file transfer mechanism.

- **Information Report Function.** A network element should have the capability of reporting information autonomously. This will be done when some information on the state or operation of the system has changed. For appropriate events in the system, a network element must be able to identify the notification as an alarm and be able to indicate the severity and cause of the condition in the report. Notifications may be logged locally. Logged notifications may be requested to be transferred using the defined file transfer mechanism.

- **Response/Report Control Function.** For responses to information requests and for information reports, it should be possible for the network operator to specify where and when the information should be sent. The operations system and network element should have the capability to configure the response/reporting to meet the following requirements:

  ✗ Information forwarding must be able to be enabled and disabled,

  ✗ Information must be forwarded to the operations system as soon as it is available,

  ✗ Information must be able to be directed to any of the various operations systems,

  ✗ Information must be able to be logged locally by the network element and optionally by the operations system, and

  ✗ Information must be retrieved from logs using appropriate filtering specifications.

11.8 FAULT MANAGEMENT

Fault management consists of a set of functions that enable the detection, isolation, and collection of an abnormal operation of a network element reported to an operations system. Fault management uses surveillance data collected at one or more network elements. The collection and reporting of the surveillance data are controlled by a set of generic functions. Fault management functions for operations system/network element interfaces deal with maintaining and examining error logs, reporting error conditions, and localizing and tracing faults by conducting diagnostic tests. The fault management supports the following management service components:

- **Alarm Surveillance.** This service deals with managing information in a centralized environment about service affecting performance degradation. Alarm functions are used to monitor or interrogate network elements about events or conditions.

- **Fault Localization or Identification.** This service requires that the management system have capabilities for determining which unit is at fault. The repair or replacement of the faulty unit can then be done to restore normal operation of the system. Such repair or replacement may be accomplished automatically by the agent, by the operations system, or manually by craft. The first step in this process is to identify the faulty unit as part of an alarm notification or test result. The alarm notifications and test results should contain identification of the repairable/replaceable unit whenever possible. Even though it is not possible to identify a specific repairable/replaceable unit, a list of potential faulty units should be provided so that additional diagnostic tests may be carried out to further localize the fault. For alarm notifications, the first localization information is provided by the identification of the object instance reporting the alarm. Second, the alarm "Probable Cause" and, optionally, "Specific Problem" values will provide the wireless network provider or manufacturer with the information that will help to localize the fault to a specific replaceable/repairable unit.

- **Fault Restoration, Correction, or Recovery.** The correction of faults should be either through repair of equipment, replacement of equipment, software recovery, logic, or cessation of abnormal conditions. If a network element uses an automatic recovery action, the management system should support the capability of the network element to notify the operations system of the changes that were made. Automatic recovery mechanisms should take system optimization into account. The operator should be able to shut down and lock out a resource before taking any fault management action. The operations system sends a request to the network element indicating the network element resources to be shut down. The network element locks out the indicated resource immediately, if it is not active. If the specified resource is active, then the lock-out is delayed until all activities are ended or until a lock-out is requested by the operations system. The fault management system should also support the establishment and selection of previous versions of software and databases (e.g., backup, and fall-back) in order to support recovery actions.

- **Testing.** Periodic scheduled and on-demand tests are used to detect faults in the system. On-demand tests are also used to assist in localizing a fault to one or more repairable/replaceable units. On-demand tests are also conducted to verify a replaceable unit before placing it into service. In-service tests may be used to gather information related to a particular wireless subscriber's activities within a network. Such data might include information related to successful and failed registrations, call attempts, handoffs, and the like. Records of these activities or events may need to be correlated across network boundaries in order to diagnose problems related to delivery of the service. Testing of wireless components should be automated to the extent that human interface in the testing process is minimized. The testing capabilities should support the running of a single test or multiple tests in parallel. Control capabilities are required to simulate the same type of test to be run with different control parameters on the same or on different system units. Once started, tests must run to completion. For controllable tests, it should be possible to monitor the operation, to suspend and resume the operation, as well as to terminate the tests either gracefully or abruptly. General requirements for tests and test management can be found in ITU X.737 and ITU X.745 specifications.

- **System Monitoring and Fault Detection.** The fault management of the network element requires capabilities to support the recognition and reporting of abnormal conditions and events. Detection also addresses activities such as trend analysis, performance analysis, and periodic testing. Fault management requires that the operations systems have a consistent view of the current state and configuration of the system that is being managed. The management system should be able to request current information about the state and configuration of the system in the form of solicited reports. The local OAM&P activities should be reported to the operations system as soon as practical. Whenever possible, a network element should generate a single notification for a single fault. A network element should support the local storage of fault information. This stored information should be accessible and erasable from the operations system. Network elements should support capabilities to allow the request and reporting of current alarm status information. This should indicate units with alarms outstanding and their severity. Fault management should have the capability to report alarms to identify failures in equipment (hardware and/or software), databases, or environmental conditions. Information about faults should be stored in the operations system for statistical verification/estimation of equipment reliability. The alarm-reporting functions are described in ITU X.733. They fall into four general areas: the generation of an event notification by a managed resource; the forwarding of that event notification to a management application; storage (temporary) of the event record; and alarm status monitoring capability.

11.9 PERFORMANCE MANAGEMENT

Performance management functions deal with a continuous process of data collection about the grade of service, traffic flow, and utilizations of the network elements. It does not affect the service provided to the wireless customer. Performance monitoring is designed to measure the overall service quality to detect service deterioration due to faults or to planning and provisioning errors. Performance monitoring may also be designed to detect characteristic signal patterns before signal quality has dropped below an acceptable level. The performance monitoring is sensitive to the sampling scheme of the monitored signal and/or choice of signal parameters calculated from the raw signal data.

Wireless network system behavior requires that performance data be collected and recorded by their network elements according to the schedule established by the operations system. This aspect of the management environment is referred to as performance management. The purpose of the *performance management* is to collect data that can be used to evaluate the operation of the system to verify that it is within the defined QoS limits and to localize potential problems as early as possible. Data are required to be produced by the network elements to support the following areas of performance evaluation:

- Traffic levels within the wireless network, including the level of both user traffic and signaling traffic;
- Verification of network configuration;
- Resource access measurements;
- Quality of service (e.g., delays during call setup); and
- Resource availability.

The production of the measurement data by the network elements should be administered by the operations system. Phases of administration of performance measurements are

- The management of the performance measurement collection process;

- The generation of performance measurement results;
- The local storage of measurement results in the network element;
- The transfer of measurement results from network element to an operations system; and
- The storage, preparation, and presentation of results to the craft.

11.9.1 Requirements on Types of Data

Typical requirements for the types of performance data to be produced by the network elements of a wireless network are discussed next.

Traffic Measurement Data. Traffic measurement data provide the information from which the planning and operation of the network can be carried out. They include

- Traffic load on radio interface (signaling and user traffic), where measured values may include pages per location area per hour, busy hour call attempts, and handoffs per hour;
- Usage of resources within the network nodes; and
- User activation and use of supplementary services.

Network Configuration Evaluation Data. Once a network plan or changes to a network plan have been implemented, it is important to be able to evaluate the effectiveness of the plan or planned changes. The measurements required to support this activity will indicate the traffic levels with particular relevance to the way the traffic uses the network.

Resource Access Data. For accurate evaluation of resource access, each count must be produced for regular time intervals across the network or for a comparable part of the network.

Quality of Service Data. QoS data indicate the wireless network performance expected to be experienced by the user.

Resource Availability Data. The availability performance is dependent on the defined objectives (i.e., the availability performance activities carried out during the different phases of the life cycle of the system) and on the physical and administrative conditions.

11.9.2 Measurement Administration Requirements

Measurement administration functions allow the system operator to manage measurement data collection and forwarding to an operations system.

Measurement Job Administration. The processes to accumulate the data and assemble them for collection and/or inspection should be scheduled for the period or periods for which collection of data should be performed. The administration of measurement job consists of the following actions:

- Create/delete a measurement job;
- Modify the characteristics of a measurement job;
- Define measurement job scheduling;
- Report and route results (to one or more operations systems);
- Suspend/resume an active measurement job; and
- Retrieve information related to measurement jobs.

Measurement Result Collection Method. The measurement data can be collected in each network element of the wireless network in a number of ways:

- Cumulative incremental counters triggered by the occurrence of the measured event;
- Status inspection (i.e., a mechanism for high-frequency sampling of internal counters at predefined rates);
- Gauges (i.e., high value mark, low value mark); and
- Discrete event registration, where data related to a particular event are captured.

Local Storage of Results at the Network Element. It should be possible for the network element to retain measurement data it has produced until they are retrieved by the operations system. These data should be retained at the network element as an explicit request from the operations system. The storage capacity and the duration for which the data should be retained at the network element will depend upon the network design.

Measurement Result Transfer. The results of the measurement job can be forwarded to the operations system when available or be stored in the network element and retrieved by the operations system when required. The measurement result can be retrieved from the network element by the operations system on request. The network element should return the current value of the measurement job with any related information but should not affect the scheduled execution of this or any other measurement jobs actively reporting that same data item.

11.9.3 Requirement on Measurement Definition

The measurements defined for the wireless network are collected at the network elements. Each network element plays its own role in the provision of the wireless telephony service and has a different perspective on the performance of the network. The measurement definition should contain a description of the intended result of measurement in terms of what is being measured.

The definition of a measurement should accurately reflect which types of events are to be included in the collection of data. If a general event description can be characterized by several subtypes, then the measurement definition should be precise as to which subtypes are included or specifically excluded from the measurement.

In a multivendor network, it is important to ensure that measurement data produced by a network element from one supplier is equivalent to the measurement data being produced by the equivalent network element from another supplier. This is particularly important when analyzing data across the whole network.

In a complex wireless network, it is easy to generate large amounts of performance data. It is essential that all data are recognizable with respect to each request.

11.9.4 Measurement Job Requirements

The measurement schedule specifies the time frame during which the measurement job will be active. The system should support a job start time of up to at least 90 days from the job creation date. If no start time is specified, the measurement job should become active immediately. The measurement job remains active until the stop time (if supplied in the schedule) is reached. If no job stop time is specified, the measurement job should run indefinitely and can be stopped only by manual intervention. The time frame defined by the measurement schedule may contain one or more recording intervals. These recording intervals may repeat on a daily and/or weekly basis and specify the time periods during which the measurement data are collected with the network element.

The granularity period is the time between the initiation of two successive gatherings of measurement data. Required values for the granularity period are 5, 15, 30, and 60 minutes. The minimum granularity period is 5 minutes in most cases, but for some measurements it may be necessary to collect data for a larger granularity period. The granularity period should be synchronized on the full hour and its value not changed during the lifetime of the job.

Scheduled measurement reports are generated at the end of each granularity period. All reports generated by a particular measurement job should have the same layout and contain the information requested by the system operator. The information may contain:

- An identification of the measurement job that generated the report;
- An identification of the involved measurement type(s) and measured network resource;
- A time stamp, referring to the end of the granularity period;
- The result value and indication of the validity for each measurement type; and
- An indication that the scan is not complete and the reason why the scan could not be completed.

Scheduled measurement reports generated at the end of each granularity period, if the measurement job is not suspended, can be transferred to the operations system in two ways:

- The reports are automatically forwarded to the operations system at the end of the granularity period (e.g., immediate notifications).
- The reports are stored locally in the network element, where they can be retrieved when required.

11.9.5 Performance Measurement Areas

The areas of performance measurement follow:

- **Traffic and Signaling Data Collection in Network Elements,** (e.g., the base station, mobile switching center, and home location register).
- **Quality of Service Data Collection.** QoS data are collected for dropped calls, connection establishment, connection quality for mobile services, connection quality for data services, call/event record integrity, and quantity per network management. For QoS, the following measurements should be taken:
  - ✗ Delay to provide a new service;
  - ✗ Delay to change subscriber data;
  - ✗ Operator complaints about user type, service type, call type, and the like;

✗ Delay to re-establish a service; and

✗ Customer service promptness.

- **Availability Measurements.**
- **Network Performance Measurements.** The following measurements should be used:

  ✗ Percentage overflow (%OFL) (count used whenever the cell site cannot allocate a traffic channel for setting up a call),

  ✗ Answer seizure ratio (ASR),

  ✗ Answer bid ratio (ABR),

  ✗ Mean holding time per seizure, and

  ✗ Abnormal termination ratio.

- **Call Destinations.** The MSC should be able to monitor call destinations as country codes, area codes, exchange codes, or any combination of them. The entities that should be measured on call destinations for network management purposes are

  ✗ Attempts per destination per hour,

  ✗ Seizures per destination per hour,

  ✗ Answers per destination per hour, and

  ✗ Count of calls affected by network management control (controls on destination).

- **Radio Interface Traffic Measurement.** The air interface is the most traffic-sensitive area in the mobile network. It is important that the maximum feasible number of measurements on the radio interface be taken. The following measurements are recommended:

  ✗ Amount of signaling leading to unsuccessful events,

  ✗ Congestion probability on mobile traffic channels,

  ✗ Lost call probability due to handoff failure,

  ✗ Call re-establishment probability, and

  ✗ Loads on the air interface.

- **Processing Throughput Measurements.** The level of processing throughput in terms of busy hour call attempts or transaction rates versus delay times should be monitored on the MSC, BS, HLR, and SS7 signaling links to allow for overload criteria to be used with appropriate network management actions to be taken. Throughput is very important in the mobile network due to extra processing required when compared to the fixed networks because of handoff, location updates, authentication, and the like.

• **Handoff Statistics.** Handoffs add to the processing and signaling load of the mobile system. From the network management viewpoint, the following data should be collected:

✘ Handoff counts per call;

✘ Success rate of handoff attempts; and

✘ Reasons for failure and retries.

In collecting the handoff data, the following should be recorded:

✘ The handoff statistics should be qualified with time of day, cells or location areas, or similar measurements, as selected by the mobile operator;

✘ Statistics concerning the handoff parameters, in order to be able to modify and tune the current handoff algorithm later on;

✘ Statistics about the causes of the handoff and their corresponding success rates;

✘ Intra-BS handoffs;

✘ Inter-BS (intra-MSC) handoffs;

✘ Inter-MSC handoffs; and

✘ Inter-mobile networks handoffs.

• **Connection Establishment and Retention.** The following data should be taken to evaluate this measurement:

✘ Percentile of ineffective calls:

(a) Mobile originated due to congestion in mobile network, congestion in attached/adjacent network, excessive delay in mobile network, excessive delay in adjacent network, premature release by calling MS during call setup, and called PSTN number not answering.

(b) Mobile terminated due to congestion in mobile network, excessive delay in mobile network, premature release by calling network during call setup, called MS busy, called MS not switched on, called MS no answer, called MS deregistration, called to barred/unobtainable MS, and failed paging to MS.

✘ Mean time of mobile network connection establishment for any service including mobile subscriber to MSC, mobile subscriber to gateway MSC, MSC to mobile subscriber, or gateway MSC to mobile subscriber.

✗ Post-dialing delays for national calls for both same mobile network operator and more than one mobile network operator case and international calls per destination country and mobile network operators involved.

✗ Mean time of service interruptions due to faults.

✗ Mean time between failures causing interruptions.

• **Connection Quality Measurements.** To evaluate connection quality, the following measurements should be taken:

✗ Probability of unintelligibility;

✗ Duration of interruption to call;

✗ Bit-error rate;

✗ Throughput for nontransparent services;

✗ Coverage for nontransparent services;

✗ Coverage for transparent services; and

✗ Error-free seconds.

• **Billing integrity measurements.**

✗ Number of billing errors (percent of bills sent);

✗ Percentile of justified complaints about billing; and

✗ Percentile of calls with tariffing errors.

11.10 SUMMARY

In this chapter, we discussed the management goals for PCS networks and provided the necessary requirements for PCS network management. We then discussed important features of the Telecommunications Management Network and outlined the five management functions: accounting management, security management, configuration management, fault management, and performance management. We also discussed the layered architecture for network management and presented the functional and physical architecture for TMN. We concluded the chapter by outlining the requirements in the five functional areas of management for a wireless network.

11.11 REFERENCES

1. ANSI T1.210, "Operations, Administration, Maintenance and Provisioning (OAM&P)— Principles of Functions, Architecture and Protocols for Telecommunication Management Network (TMN) Interfaces."

2. ANSI T1.215, "Operations, Administration, Maintenance and Provisioning (OAM&P)—
    Fault Management Messages for Interfaces between Operations Systems and Network
    Elements."

3. ANSI T1.227, "Operations, Administration, Maintenance and Provisioning (OAM&P)—
    Extension to Generic Network Model for Interfaces between Operations Systems
    across Jurisdictional Boundaries to Support Fault Management–Trouble Administra-
    tion."

4. "Draft ANSI T1.XXX—1996, American National Standard for Telecommunications—
    Operation, Administration, Maintenance, and Provisioning (OAM&P)—Performance
    Management Functional Area Services for Interfaces between Operations Systems
    and Network Elements."

5. "Draft ANSI T1.XXX—1996, American National Standard for Telecommunications—
    Operation, Administration, Maintenance, and Provisioning (OAM&P)—Technical
    Report on PCS Accounting Management Guidelines," T1M1.5/95-011R4.

6. ITU-T Rec. M.3010, "Principles for a Telecommunication Management Network
    (TMN)," Draft 950630, ITU—Telecommunications Standardization Sector, 1995.

7. CCITT Rec. M.3400, TMN Management Functions, ITU—Telecommunications Stan-
    dardization Sector, 1992 (under revision, December 1996).

8. Draft CTIA, "Requirements for Wireless Network OAM&P Standards," CTIA OAM&P
    SG/95.11.28.

9. EIA/TIA IS-124, "Cellular Radio Telecommunications Intersystem Non-signaling Data
    Message Handler (DMH)."

10. Garg, V. K., and Wilkes, J. E., *Wireless and Personal Communications System,* Pren-
    tice Hall, Upper Saddle River, NJ, 1996.

Interconnection Between Systems

12.1 INTRODUCTION

CDMA systems do not exist in a vacuum. They must coexist with other wireless systems that share the same frequency spectrum, and they must be interconnected to other wireless and wireline voice and data networks. In this chapter, we will examine the interworking issues between different systems in sufficient detail to examine various dual-mode technologies being built and proposed. We will then discuss the dual-mode cellular/PCS CDMA and AMPS systems and phones that are being proposed by TIA committees TR-45 and TR-46. Although there is no active standards work in the area of dual-mode digital systems, we will identify the issues that need to be resolved to construct the systems and phones. The key to service interoperability is a seamless connection between the wireless and wireline networks. The work on intelligent networks for the wireline network has been extended to the wireless arena; therefore, we will conclude the chapter with a discussion of Wireless Intelligent Networking being planned by the international standards community.

12.2 INTERWORKING ISSUES

The interworking between a wireless telephone and the worldwide wireline network is a topic that is being actively worked throughout the world. Early mobile telephone users often had to dial special access codes or use dialing sequences that were different from the wireline network. While never optimum, it was acceptable since mobile telephony was a premium service with a low number of users. As wireless telephony becomes ubiquitous, service providers and telephone stores can no longer afford the special training and support necessary to teach new users how to use the phone. Therefore, new wireless phones must operate identically to the wireline network.[1] In this section, we will discuss the issues related to obtaining seamless operation between the wireline and wireless networks. We will discuss only the issues here. A more complete treatment of issues and potential solutions can be found in the book *Wireless and Personal Communications* [3] and by actively following the activities of the standards bodies in the United States, Europe, and the world.

The wireline telephone network is primarily based on analog telephones and analog local loops using dual-tone multifrequency (DTMF) signaling between the subscriber and the central office and digital switching with digital signaling between central offices. Recently, business and residential subscribers are switching to Integrated Service Digital Network (ISDN) with digital telephones and digital signaling. Neither of these two methods currently uses wireless communications. Although most cellular telephones use analog frequency modulation (FM), the signaling is digital since DTMF tones do not have high reliability when sent over a fading radio channel. While there are clearly digital wireless phones, most do not support an ISDN channel with two 64-kbps bearer channels and a 16-kbps signaling channel. Thus, the voice and signaling protocols used in a wireless system must be converted into the protocols used on the wireline network. The base station or the mobile switching center will perform the conversion, called *interworking*.

Low-speed data are transmitted over the telephone network today using voice band modems that range from 1200 to 28,800 baud. Some of the uses for this data are

- Accessing electronic mail,
- Accessing remote computers,
- Transferring files,
- Using facsimile transmissions, and
- Performing transaction services (e.g., credit card validation).

---

1. It is unlikely that new wireless phones will generate a "dial tone" when off-hook and process dialed digits using the same procedures as the wireline network of today. However, new wireline digital services and the proliferation of advanced calling features may result in the two services moving to a common mode of operation.

Transmission at higher rates is typically used for video conferencing, mainframe-to-mainframe communications, and other uses that would not initially be carried over a wireless mobile link.

Except for those PCS supporting pulse code modulation or adaptive differential pulse code modulation (e.g., W-CDMA), the speech-coding systems for transmission of voice have been optimized for the voice transmission and have not been optimized for the transmission of non-voice signals such as voice band modems. Therefore, both the mobile station and the cellular/personal communications system must have interworking capabilities to provide the wide range of services currently using voice band modems. Over the air interface, the data must be transmitted digitally since that is the only option available. If the air interface supports a data rate higher than the basic data rate of the voice band modem, then interworking is possible. Interworking is not possible if the required data rate is higher than the needed data rate. For example, 28.8-kbps data service is not possible over a 16-kbps data link. Those systems supporting PCM or ADPCM can interwork with voice band modems without a data interworking platform. However, error performance may suffer because of the high error rate of the radio channel.

The various telephone administrations and telephone companies around the world have adopted a variety of national dialing plans for reaching telephones in their nations. They have assigned a (nationally) unique number to each telephone in their administration. In North America, there is a uniform 10-digit dialing plan. In other parts of the world, the numbering plan permits numbers of variable lengths from 6 to 15 digits. Even though the differences between numbering plans can easily be handled by the wireline telecommunications network, it creates complications for roaming mobile telephones. A telephone number that is unique in one nation may not be unique worldwide. Thus a mobile telephone registering in a wireless system in other than its home country may identify itself to the network with a number that duplicates a number from another country. The various standards bodies are solving this problem by migrating all mobile telephone numbers to a 15-digit International Mobile Station Identification (IMSI) number. The lack of a unique identity will exist until all current mobile telephones numbers are upgraded to the IMSI (See chapter 4).

In chapter 7, we studied the call flows for a variety of telephony functions. These call flows, while specific to CDMA, are part of a general set of call-processing functions based on the North American cellular/PCS network. They are part of a set of requirements for services [2], intersystem communications [3], base station-to-mobile switching center communications [4], and an air interface [5, 6]. The overall system is based on ISDN and SS7 signaling with overlays for mobile communications. The overlay is called a mobile application part (MAP). We have examined the IS-41-based MAP, but there is a second MAP used in Europe and migrating to North America that is based on the global system for mobile communication (GSM) system and called the GSM MAP. The overall call processing is done somewhat differently between the MAPs, and the security mechanisms are different. Significant work remains to provide interworking between these two MAPs.

A critical feature for wireless systems where the user roams to other systems is the ability to bill for calls placed or received on a visited system. A critical component to this is to support a common billing record system across all systems. In North America, the TIA IS-124 standard provides the common billing format. In other areas of the world, work is under progress to generate a common billing standard.

The final area where interworking is important is support for multiple air interfaces. The dominant interfaces in North America will be AMPS, CDMA, TDMA, and GSM. Methods must be developed for interworking between these and any other interfaces (e.g., PACS). The next two sections are devoted to these issues.

12.3 DUAL-MODE DIGITAL/AMPS SYSTEMS AND PHONES

In the United States, a decision by the Federal Communications Commission to require one nationwide standard, and thus support roaming between any systems in the country, resulted in the rapid deployment of cellular systems. In Europe, several nations built their own systems that were incompatible with systems in other European nations. Similarly, Japan built its own cellular system. Thus, throughout the world, there are at least five different, incompatible, first-generation cellular standards. Each of these standards depends on frequency modulation of analog signals for speech transmission, out-of-band signaling for call setup, and in-band signaling to send control information between a mobile station and the rest of the network during a call.

When the GSM system was designed for Europe, there was no push for a dual-mode (analog/digital) telephone since the analog phone would work only in one country. Thus, there was a push for a new Europe-wide digital system that was rapidly deployed. Unlike the incompatible analog systems, customers of the new system would gain roaming capabilities throughout Europe.

In North America, the EIA IS-553 cellular standard (now TIA IS-91) is fully deployed, and customers enjoy roaming throughout Mexico, the United States, and Canada. Any new digital system would not be fully deployed in the early years and, therefore, would not be competitively viable. The North American standards permit coexistence with the first-generation standard, the advanced mobile phone system (AMPS), and add a digital voice transmission capability for new digital equipment. Channels in a given geographic area may be assigned to either digital or analog transmission, whereas in another area the two signals may share channels. The North American TDMA Standard, IS-54/IS-136, enhances, rather than replaces, the analog cellular technology. Similarly for CDMA, a dual-mode design is possible; however, as we discussed in chapter 9, for the first CDMA channel, nine analog channels at each base station must be devoted to CDMA with an overall improvement in spectrum efficiency. Additional CDMA channels are obtained by removing six analog channels from service from each base station.

When a dual-mode phone and system is introduced, the designers must solve the following problems:

- Spectrum sharing between the analog and digital systems,
- Phone initialization,
- Handoffs from analog to digital and from digital to analog,
- Voice privacy, and
- Compatible call control between analog and digital operation.

Chapter 9 discusses the spectrum sharing between analog AMPS and CDMA. We describe the growth process from a purely analog system to a mixed analog/digital system to a pure CDMA digital system.

The IS-95 standard defines the operations of the mobile station during initialization. The MS must chose the frequency band to use, A or B band, and the modulation to use, analog or CDMA system. The frequency band options can be chosen independently of the modulation options. The preferences are stored in the MS when service is activated and often can be changed by the user of the phone. Some typical preferences (as defined by the standards) are

- System A only,
- System B only,
- System A preferred (if No Service can choose System B),
- System B preferred (if No Service can choose System A),

- CDMA only,
- Analog only,
- CDMA preferred (if No Service can choose Analog operation), and
- Analog preferred (if No Service can choose CDMA operation).

Other preferences that are often designed into the MS are

- Home only,
- No service on a list of SIDs, and
- Chose alternate telephone number if roaming and alternate telephone number has same SID as the roaming system.

All these options are for a cellular dual-mode MS. When a dual-mode, dual-frequency cellular/PCS AMPS/CDMA phone is standardized, PCS frequencies will be added to the CDMA options.

When CDMA service is available, the dual-mode MS would normally start a call on a CDMA channel. The analog channels will be reserved for analog-only MSs. As the MS moves from one area of the system to another, it will be handed off between base stations. At some point, it may move outside the coverage area of the CDMA base stations. When that event happens, the system will process a hard handoff to an analog cellular channel. The handoff to the analog channel has several consequences. The call will continue as an analog call until it is ended. The standards do not support the handoff back to a CDMA channel. The message to inform the MS of a new CDMA channel, using the analog blank-and-burst data transmission, is excessive and would result in a long handoff outage. Also, unless other means are provided, the user has lost voice privacy since analog FM signals are easily intercepted on older police/fire scanners or older TV sets that cover channels 70–83.

The IS-95 standard supports analog and CDMA operation as separate functions once the MS is initialized. A handoff to the analog operation is allowed to change the phone from one mode to the other. Since the two modes are independent, it would be possible for the call-processing functions to be different in the two modes. However, cellular and personal communications systems designed for use in North America use a common intersystem communications functionality defined in the IS-41 protocol. Thus, the call-processing functions will be the same for either digital or analog operation.

Other dual-mode analog/digital systems are either defined or being defined by the standards bodies. An AMPS/TDMA MS is defined in IS-54 and IS-136 and provides similar functionality to the AMPS/CDMA MS. The European GSM system has been modified for use in North America and will see deployment in many cities. Therefore, a dual-mode AMPS/GSM MS will also be defined. Unfortunately, the operation of the MS may not be as seamless as a dual-mode AMPS/CDMA MS since the GSM MAP is different from the IS-41 MAP and telephony services may operate differently in different systems (AMPS and GSM).

12.4 DUAL-MODE DIGITAL SYSTEMS

In this section, we will examine the operation of a dual-mode digital system. No dual-mode systems currently exist, so we will limit our discussion to the problems that must be solved to build them. We will first examine the issues of the mobile station and then examine the system issues.

Two key problems need to be solved for a dual- (or multiple) mode phone: initial system selection and handoffs. When a phone is first turned on, it goes through a sequence of events that initializes the phone and registers it with a system. Although we can force a set of preferences similar to those described in the previous section, a multimode phone needs to have a set of procedures that will enable it to try frequencies and protocols for two or more air interfaces. Dual digital mode MSs may require multiple receivers and transmitters and extensive software to control them. The start-up sequence may be lengthy because the phone does not know what system it is in when first powered up. The user of the phone may help the phone by giving information about location but, in general, the phone should do its own analysis. Users of a dual mode system roaming into new areas may see start-up delays of several tens of seconds or more before service is available. After the phone has registered with a system, calls can be placed or received.

After a call is established, it may be necessary to handoff the call to another radio system. All air interfaces support handoff; and the issues of handoff between radio systems (using the same air interface) connected to the same or a different PCS switch are well understood and have been discussed in chapter 7.

Handoffs between systems essentially require transmission and signaling facilities between the switches of each system. Other than the extra facilities, the handoff between two radio systems on two different systems (using the same air interface) does not differ from the handoff between radio systems on the same PCS or cellular system.

However, when the air interfaces differ between the two radio systems, then the problem is more complicated. All the digital systems assume that the data receiver can quickly regain synchronization when a call is handed off to another base station. In particular, when CDMA is used, the receiver in the MS receives both radio systems simultaneously and uses signals from both to produce a composite signal. When the handoff occurs between two different air interfaces, new handoff messages will be necessary so that the MS can be informed about the characteristics of the new systems. A second receiver may also be necessary so that the MS can be listening to the new radio system before the handoff occurs. Otherwise, the handoff delay may be excessive and calls may be dropped.

If the handoff algorithm differs (MS initiated, anchor switch initiated, or anchor radio system initiated), then new handoff procedures and messages need to be developed and tested.

Clearly extensive modifications to the MS and the network will be needed to support handoffs between different air interfaces in different systems.

For good performance on MS initialization, each system in an area may need to broadcast information about all systems in the area. Thus, when an MS enters a new area, if it can receive any transmissions from any system, it can find its preferred system. System information messages that inform MSs of other systems do not currently exist except for limited case of public/private systems of the same modulation type. All air interface protocols need upgraded capabilities on the system information channel to support multimode digital terminals.

The alternative would require that the MS try multiple sets of frequencies and protocols until it starts decoding information, which will result in a lengthy start-up process and time.

For dual-mode CDMA/GSM MSs, an additional problem results and is similar to the dual-mode AMPS/GSM problem. CDMA supports the IS-41 MAP which is different from the GSM MAP. Thus, a dual-mode CDMA/GSM MS will operate differently depending on which system it is being used on. This may make the user unwilling to place calls in areas where CDMA service is not available and defeats the purpose of the dual-mode phone.

12.5 WIRELESS INTELLIGENT NETWORKS

For the wireline network, BellCore has defined a set of protocols called intelligent network (IN) [1] that enable a rich set of new telephony capabilities to be generated without additional software development in the central switch. As IN has grown in popularity for the wireline network, the wireless network is also embracing the concept. IN improves the ability to locate and efficiently direct calls to roaming MSs and provides other advanced features. In this section, we will examine the work of the international standards bodies and TR-45/TR-46 in the United States to add IN to the wireless systems.

The International Telecommunications Union (ITU) has adapted IN for use in wireless networks. Question 8 of Study Group 11 has defined a communications control plane (fig. 12.1) and a radio resource control plane (fig. 12.2) using IN concepts.

On the network side of the communications control plane, the following functional elements are defined:

- The **Bearer Control Function (BCF)** provides those bearer functions needed to process handoffs. A conference bridge to support soft handoffs is a common example.
- The **Bearer Control Function for the Radio Bearer (BCFr)** provides the functions necessary to select bearer functions and radio resources. It also detects and responds to pages from the network and performs handoff processing. Some example bearer func-

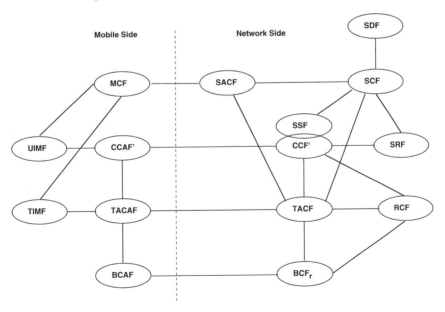

Figure 12.1 Communications control plane.

tions are PCM voice, ADPCM voice, packet data, and circuit-switched data.

- The **Call Control Function-Enhanced (CCF')** provides the call and connection control in the network. Some examples include establishing, maintaining, and releasing call instances requested by the CCAF', providing IN triggers to the service switching function (SSF), and controling bearer connection elements in the network.
- The **Service Access Control Function (SACF)** provides the network side of mobility management functions. Some examples are registration and location of the mobile station.
- The **Service Control Function (SCF)** contains the service and mobility control logic and call processing to support the functions of a mobile terminal.
- The **Service Data Function (SDF)** provides data storage and data access in support of mobility management and security data for the network.
- The **Specialized Resource Function (SRF)** provides the specialized functions needed to support execution of IN services. Some examples are dialed-digit receivers, conference bridges, and announcement generators.
- The **Service Switching Function (SSF)** provides the functions required for interaction between the CCF' and the SCF. It supports extensions of the CCF' logic to recognize IN triggers and interact with the SCF. It manages the signaling between the CCF' and the SCF and modifies functions in the CCF' to process IN services under control of the SCF.
- The **Terminal Access Control Function (TACF)** provides control of the connection between the mobile station and the network. It provides paging of mobile stations, page response handling, handoff decision, completion, and trigger access to IN functionality.

On the mobile side, the following functional elements are defined:

- The **Bearer Control Agent Function (BACF)** establishes, maintains, modifies, and releases bearer connections between the mobile station and the network.
- The **Call Control Agent Function-Enhanced (CCAF')** supports the call-processing functions of the mobile station.
- The **Mobile Control Function (MCF)** supports the mobility management functions of the mobile station.

- The **Terminal Access Control Agent Function (TACAF)** provides the functions necessary to select bearer functions and radio resources. It also detects and responds to pages from the network and performs handoff processing.
- The **Terminal Identification Management Function (TIMF)** stores the terminal-related security information. It provides terminal identification to other functional elements and provides the terminal authentication and cryptographic calculations.
- The **User Identification Management Function (UIMF)** provides user-related security information similar to the TIMF.

Both the TIMF and the UIMF can be stored in either the mobile station or a separate security module often implemented in a smart card.

The radio resource control plane (fig. 12.2) is in charge of assigning and supervising radio resources. Four function entities (two on the mobile side and two on the network side) perform the functions of the radio access subsystem:

- The **Radio Resource Control (RRC)** provides functionality in the network to select radio resources (e.g., channels and spreading codes), make handoff decisions, control the RF power of the mobile station, and provide system information broadcasting.
- The **Radio Frequency Transmission and Reception (RFTR)** provides the network side of the radio channel. It provides the radio channel encryption and decryption (if used), channel quality estimation (data-error rates for digital channels), sets the RF power of the mobile station, and detects accesses of the system by the mobile station.

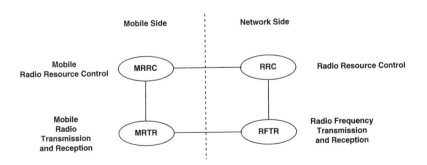

Figure 12.2 Radio resource control plane.

- The **Mobile Radio Resource Control (MRRC)** processes the mobile side of the radio resource selection. It provides base station selection during start-up, mobile-assisted handoff control, and system access control.
- The **Mobile Radio Transmission and Reception (MRTR)** provides the mobile side of the radio channel and performs functions similar to the RFTR.

The two control planes interact to provide services to the mobile station and the network.

The TR-46 PCS network reference model working group has generated a simplified version of the ITU model and has explicitly shown the operations functions on the model. The IN function reference model for PCS (fig. 12.3) has many of the functional elements with the same element name as the ITU model. The differences follow:

- The **Radio Terminal Function (RTF)** contains all functionality of the mobile side of the reference model.
- The **Radio Access Control Function (RACF)** is similar to the SACF and provides mobility management functions.

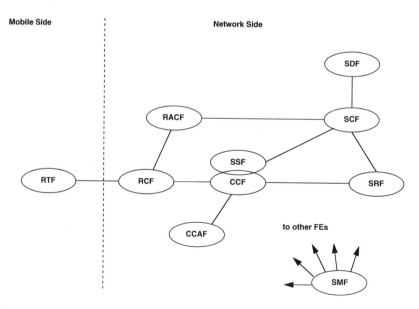

Figure 12.3 IN functional reference model for PCS.

- The **Radio Control Functions (RCF)** provide the capabilities of the TACF, the BCFr, and the BCF in the ITU model. It provides the radio ports and the radio port controller capabilities in the PCS network.
- The **Call Control Agent Function (CCAF)** provides access to the wireless network by wireline users.
- The **Service Management Function (SMF)** provides the network management functions for each functional element.

With both the ITU and the TR-46 reference models, the functionality to support all the features and capabilities for a wireless network can be partitioned into different functional elements and still meet the variety of national and worldwide standards. Therefore, no exact partitioning of the functions of the network reference models described in chapter 4 can be made. We encourage you to examine the standards and implementations of the various manufacturers for example partitioning.

12.6 SUMMARY

In this chapter, we discussed the interconnection of a wireless network to the worldwide wireline telecommunications network. We showed that, for both voice and data, the wireless network must perform conversions to enable calls to proceed onto the wireline network. We demonstrated that wireless telephones roaming to different areas of the world present issues of numbering and billing that do not occur on a wireline network. The use of a standard 15-digit International Mobile Station Identification will solve the numbering problem. Only the support of a standard billing format that can be quickly sent between different areas of the world will solve the billing problem.

We then described the operation of the dual-mode AMPS/CDMA MS as it provides service to users on either the analog or digital system. Since there are no dual-mode digital systems, we examined the problems that must be solved by the design of the systems and phones but await the standards bodies to offer the solutions. Clearly custom solutions will be provided before the standards are written and adopted, but these custom solutions may not work in all systems.

Finally we examined the migration of the intelligent network concepts from the wireline network to the wireless network. As with many of the issues raised in this book, we encourage you to follow the worldwide standards activities to obtain the latest plans to solve the issues discussed in this chapter.

12.7 REFERENCES

1. BellCore, "Advanced Intelligent Network (AIN) Release 1 Switching System Generic Requirements," TA-NWT-001123, Issue 1, May 1991.
2. ITU Study Group 11, "Version 1.1.0 of Draft New Recommendation Q.FNA, Network Functional Model for FPLMTS," Document Q8/TYO-50, September 15, 1995.
3. Garg, V. K., and Wilkes, J. E., *Wireless and Personal Communications Systems*, Prentice Hall, Upper Saddle River, NJ, 1996.
4. TIA Interim Standard, IS-41C, "Cellular Radio Telecommunications Intersystem Operations."
5. TIA Interim Standard, IS-54B, "Cellular System Dual-Mode Mobile Station–Base Station Compatibility Standard."
6. TIA Interim Standard, IS-91, "Cellular System Mobile Station–Base Station Compatibility Standard."
7. TIA Interim Standard, IS-95, "Mobile Station–Base Station Compatibility Standard for Dual-Mode Wideband Spread Spectrum Cellular System."
8. TIA Interim Standard, IS-136, "Cellular System Dual-Mode Mobile Station–Base Station Compatibility Standard."
9. TR-46, "Personal Communications Services Network Reference Model for 1800 MHz," PN-3436 Baseline Text Version 1, June 26, 1995.
10. EIA/TIA IS-124, "Cellular Radio Telecommunications Intersystem Nonsignaling Data Communications (DMH)."

Evolution of CDMA Technology for Wireless Communications

13.1 INTRODUCTION

In previous chapters, we discussed current standards for CDMA technology and their application to wireless communications. Presently, these standards are being implemented by manufacturers and being deployed by service providers. The wireless industry is investigating methods to improve and to reduce costs so that the mobile subscriber will experience better service at a reduced cost.

This chapter examines several areas that address these goals. First, service providers are seeking ways to reduce administrative costs by streamlining the activation procedure for new mobile subscribers. This chapter discusses over-the-air service provisioning which supports this capability. Second, advances in digital technology will make it possible to improve the quality of speech coders at a given digital rate. This chapter discusses the enhanced variable rate codec, which provides better performance than the current standardized speech processor. Third, the wireless industry is studying improvements to transmission schemes over the air interface. The resulting improvements will increase channel capacity. To mobile subscribers, this translates to a better service at a lower cost. This chapter discusses interference cancellation, multiple beam adaptive antenna arrays, and improvements of the handoff algorithm, which are three separate approaches for achieving the capacity improvement objective.

13.2 OVER-THE-AIR SERVICE PROVISIONING

Most mobile subscribers currently must obtain and initialize service through an authorized dealer rather than directly from the service provider.[1] From the point of view of the service provider, this procedure has several disadvantages:

- The service provider pays a commission to the authorized dealer. This is an expense to the service provider.
- The service provider does not have total control during this process; consequently, there is opportunity for fraud to occur.
- The mobile subscriber is inconvenienced in that the subscriber must visit the authorized dealer. Several hours may be required to complete this operation, and, furthermore, service may not be immediately activated.

Thus, the wireless industry is investigating ways to simplify the activation procedure and to reduce associated expenses. Standards body TIA TR-45 is currently developing a feature called over-the-air service provisioning (OTASP) to address this need. OTASP allows the service provider to configure a mobile station directly and to provision the mobile during a call initiated by the mobile subscriber. To initiate service, the mobile subscriber makes a mobile-originated call by dialing an activation code. The call is established between the mobile subscriber and the Customer Service Center (CSC), which is operated by the service provider. OTASP assumes that a conversation between the mobile subscriber and a CSC attendant is established so that information (e.g., credit card information and service options) can be exchanged. However, the exact procedure is determined by the service provider and may vary among service providers. As an example, the service provider may wish to reauthenticate the mobile before credit card information is exchanged. This allows the credit card information to be encypted using signaling message encryption.

OTASP defines procedures for the following operations:

1. Enable uploading the mobile's indicator data parameter block from the mobile to the CSC.

---

1. Some service providers have their own stores, but the cost of operating them is high.

2. Download the A-key to the mobile.[2]
3. Program the mobile's indicator data parameter block.
4. Reauthenticate the mobile.

The mobile's indicator data parameter block contains information such as the station class mark (SCM), SID/NID list (as discussed in chapter 3), and access overload class (ACCOLC). The indicator data parameter block consists of a subset of the mobile's number assignment module (NAM). Also, OTASP allows the service provider to download the mobile's A-key. Finally, the mobile station may be reauthenticated so that either voice privacy or signaling message encryption or both can be activated during the remainder of the call to complete the transaction with a high degree of security. The A-key, which is a 64-bit quantity, is needed to support authentication, voice privacy, and signaling message encryption.

Without OTASP, either the mobile subscriber or the authorized dealer must enter the A-key into the mobile station or the mobile station must be preprogrammed with the A-key. None of these choices is desirable. This operation is very important since manually entering the A-key has a significant probability of error and fraud. However, the A-key cannot feasibly be sent to the mobile over the air interface without a sufficient degree of encryption. It is imperative that the A-key be known at only the authentication center (AC) and the mobile station. At intermediate points between the AC and the mobile station, the A-key must be securely encrypted. The wireless industry has selected the Diffie-Hellman key exchange procedure (DHKEP) to address this concern. The DHKEP is summarized by the following steps:

1. The AC chooses a 160-bit random number a, such that $4 \leq a \leq 2^{160} - 1$.
2. The AC chooses a 160-bit random number g, such that $1 < g \leq 2^{160} - 1$.
3. The AC chooses a large prime number p having a 512-bit representation, such that $2^{511} < p \leq 2^{512} - 1$.
4. The AC sends g and p to the mobile station. Of course, this information traverses part of the network and is transmitted over the air interface. However, there is no attempt to encrypt g and p.

---

2. The A-key is a 64-bit pattern stored in the mobile station that is used to generate/update the mobile station's Shared Secret Data. Refer to section 2.3.12 (Authentication, Encryption of Signaling Information/User Data) of TIA IS-95A for further information.

5. The mobile station chooses a 160-bit random number b, such that $4 \leq b \leq 2^{160} - 1$.

6. The mobile station calculates $M = g^b \bmod p$.

7. The mobile station sends M to the AC over the air interface.

8. The AC calculates $B = g^a \bmod p$.

9. The AC sends B to the mobile station over the air interface.

10. Both the mobile station and the AC calculate $X = g^{ab} \bmod p$, where $X = M^a \bmod p$ (as calculated by the AC) and $X = B^b \bmod p$ (as calculated by the mobile station). Both the mobile station and the AC can derive the temporary A-key by truncating X to the 64 least significant bits.

11. Both the mobile station and the AC store the temporary A-key. The temporary A-key is considered the A-key only after the CSC instructs the mobile station to commit its temporary memory to semipermanent memory.

Once the temporary A-key is created at the mobile station and at the AC, the temporary shared secret data (SSD)[3] can be created. OTASP has defined a procedure so that the reauthentication can be invoked during the call. Without this new procedure, authentication requires the initiation of a new call, which means that the original call would be ended. This, of course, would present a serious complexity to the transaction. Once reauthentication is completed, voice privacy and/or signaling message encryption can be invoked. At this time, user-sensitive information can be transmitted over the air interface with a high degree of security. Once the transaction has been completed, the CSC initiates a *data commit order,* which transfers any downloaded data, the temporary A-key, and temporary SSD to semipermanent memory. At that time, the mobile station and the AC consider the programmed data as permanent.

13.3 IMPROVEMENT OF SPEECH CODERS

In order to quantize the quality of service, the wireless industry measures the speech coder's performance by the mean opinion score (MOS). The MOS is determined by statistically combining the opinions of a group of human listeners. An MOS has a range from 1 to 5, where 5 means excellent voice quality, and 1 means unsatisfactory voice quality.

---

3. The shared secret data is a 128-bit pattern that is needed for the authentication and encryption process. Refer to section 2.3.12 of TIA IS-95A and TIA PN-3569 for further information.

Toll quality has an MOS score of 4, signifying good voice quality. With analog operation, a service provider typically wishes to provide service corresponding to an MOS score of 4. The existing mobile subscriber will expect this grade of service. Thus, the grade of service provided by CDMA systems will be gauged by this quality of service.

It is anticipated that by mid 1996, standards body TIA TR-45.5 will complete the specifications for an improved variable rate speech coding algorithm for the 9.6-kbps physical layer (i.e., rate set 1) (refer to chapter 5). This is the same physical layer that the TIA IS-96A variable rate speech coder uses. Since the new speech coder is an improvement with respect to the existing 8-kbps speech coder,[4] the Telecommunications Industry Association calls the new algorithm the enhanced variable rate codec (EVRC) algorithm. This algorithm is applicable both to IS-95A mobiles (800 MHz) and to ANSI J-STD-008 mobiles (1.8–2.0 GHz). Like the TIA IS-96A algorithm, the EVRC algorithm is based upon the code-excited linear prediction (CELP) algorithm (see fig. 13.1).

The EVRC speech coder uses the relaxed code-excited linear prediction (RCELP) algorithm, and thus it does not match the original residual signal but rather a time-warped version of the original residual that conforms to a simplified pitch contour. This approach reduces the number of bits per frame that are dedicated to pitch representation. This allows additional bits to be dedicated to stochastic excitation and to channel impairment protection. The EVRC algorithm categorizes speech into full rate (8.55 kbps), half rate (4 kbps), and eighth rate (0.8 kbps) frames that

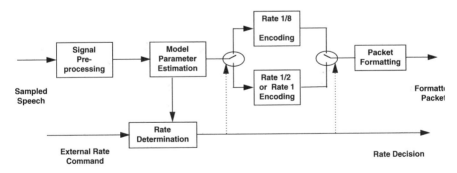

Figure 13.1 Functional diagram of the EVRC algorithm. (Reproduced under written permission of the copyright holder [TIA].)

4. Actually, the maximum data rate is 8.55 kbps, but the wireless industry references the rate as being equal to 8 kbps.

are formed every 20 ms. The TIA IS-96A algorithm categorizes speech into these rates as well as quarter rate frames.

The EVRC algorithm offers a significant performance improvement over the IS-96A speech coder (see table 13.1). These measurements do not include the effects of acoustical noise (e.g., car noise). The improvement ranges from an MOS increase of approximately 0.66 for an error-free CDMA channel to 0.95 for a channel having a frame error rate[5] of 3 percent. At a frame error rate of 1 percent, the speech coder performance of the TIA IS-96A coder is 3.17 (fair), while that of the EVRC coder is 3.83 (near toll quality). The performance at a frame error rate of 1 percent is important since channel capacity of a CDMA system is often calculated with the assumption that power control maintains the frame error rate within 1 percent. Table 13.1 also shows the performance of the CDMA Development Group's[6] 13-kbps (CDG-13 kbps) speech coder, which offers high voice quality but results in a decrease in channel capacity of approximately 40 percent. The MOS improvement of the CDG-13 kbps speech coder over the EVRC speech coder is approximately 0.12. The improvement at the frame error rate of 1 percent is not nearly as great as the improvement when transitioning from the TIA IS-96A speech coder to the EVRC speech coder. Given these observations, it is expected that the EVRC algorithm will be the basis of the speech coders in the future.

Table 13.1 Speech coder performance (MOS) as function of frame error rate

Frame Error Rate, %	CDG-13 kbps	EVRC	IS-96A
0	4.00	3.95	3.29
1	3.95	3.83	3.17
2	3.88	3.66	2.77
3	3.67	3.50	2.55

A CDMA system will need to contend with different type of speech coders (i.e., TIA IS-96A, CDG-13 kbps, and EVRC speech coders). A CDMA system may or may not support all types of these speech coders; however, the CDMA system must be able to negotiate the type of speech coder with the mobile station. As mentioned in chapter 3, this is done

---

5. The frame error rate (FER) is defined as the number of bad frames per unit of time.

6. The CDMA Development Group is a consortium of companies that developed this 13-kbps speech coding algorithm as a proprietary service option.

though service negotiation. The TIA IS-96A, CDG-13 kbps, and EVRC speech coders are assigned service option values of 1; 32,768; and 3, respectively (refer to table 3.3).

With greater penetration of personal communication services, there will be a greater importance of mobile-to-mobile calls. If both mobile stations use the same speech coder, then it is possible to send the digital output from one mobile station to the other. If this is not done, then degradation due to tandem operations of the coders occurs since two additional conversions are necessary (see fig. 13.2).

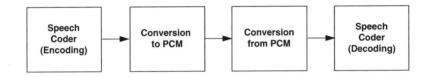

Figure 13.2 Tandem operation of speech coders.

The wireless industry is discussing this topic; however, at this time there are no definite plans to standardize a solution. At first glance, this issue seems to be a straightforward problem; however, it is not. First, there is no assurance that the speech coders at the originating and the terminating ends are the same. If they are not, then the speech coders cannot talk directly with each other. Second, a dual-mode mobile may change call modes from CDMA to analog during a call. Third, call supplementary services such as three-way calling may conference a non-CDMA party to the call. This possibility vastly increases the complexity of the conference circuit.

13.4 INTERFERENCE CANCELLATION

CDMA technology provides a significant gain in call capacity with respect to analog technology. So one might ask, "Why should the wireless industry be concerned with capacity improvements?" There are at least three reasons that motivate the wireless industry:

- If the number of mobile subscribers continue to grow at even a portion of the current rate, the assigned radio spectrum will be depleted in the future.
- New services, such as wideband data and video, are much more spectrum-intensive than the voice service.

- There is the economy of numbers. As the number of mobile sub-scribers being served by a wireless system increases, the cost per subscriber decreases.

One obvious way of increasing the capacity is to develop a reduced-rate speech coder. As discussed in the previous section, the EVRC speech coder will be introduced in the near future (probably within several years). This speech coder has a maximum data rate of 8.55 kbps, while the average data rate is approximately 4.4 kbps. In the near future, it is doubtful that variable rate speech coding technology will make a major breakthrough to reduce the average data rate appreciably, even though the voice quality will be improved at the given data rate.

Current research is investigating the practicality of canceling the interference attributed to other mobile subscribers being served by a given base station [1]. This process is a form of multiuser detection. Without such a process, interference from other mobile users translates into an increased level of noise associated with the user's CDMA channel. The base station simultaneously processes all users that are being served by that base station. Thus, this procedure has one input and multiple outputs, each output corresponding to one user. (See fig. 13.3.) The matched filter bank is implemented by multiplying the received signal by the user's spreading function $s_i(t)$. The output of the matched filter bank can be expressed in matrix notation as

$$\left[y\right] = \left[R\right] \bullet \left[W\right] \bullet \left[x\right] + \left[z\right] \qquad (13.1)$$

where $[R]$ and $[W]$ are $K \times K$ matrices,
$[y]$ is the $K \times 1$ vector representing the processed output signal,
$[x]$ is the $K \times 1$ vector representing the transmitted signal, and
$[z]$ is the $K \times 1$ vector representing the received noise.

The matrix R is given by:

$$R_{ij} = \int_0^T s_i(t) \times s_j(t)dt \qquad (13.2)$$

The matrix $[W]$ is a diagonal matrix with each diagonal element representing the channel attenuation of the corresponding mobile's signal path. The output of the matched filter bank is processed through another matrix filter that represents the inverse of $[R]$, (i.e., $[R]^{-1}$). The decision for the ith mobile's signal is calculated by the step function:

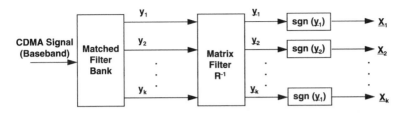

Figure 13.3 Multiuser detection.

$$\underline{x}_i = \mathrm{sgn}(\underline{y}_i) \tag{13.3}$$

This approach eliminates multiple-access interference, but it is affected by the channel noise. With the approach in figure 13.3, it is assumed that the receiver uses synchronous detection and that each mobile's signal travels along a single path. The effects of the second restriction can be reduced by incorporating a RAKE receiver as is discussed in chapter 2. The first restriction exists with a CDMA system (TIA IS-95A), since the pilot signal is not included in the reverse traffic channel. However, with W-CDMA, the pilot is transmitted on the reverse path, and thus synchronous detection is possible. Without synchronous detection, we must use noncoherent multiuser detection. More details are presented in [1].

Processing is limited to canceling interference that is associated with that base station. Otherwise, each base station would be required to send an inordinate amount of data about its associated mobile users to the other base stations. Multiuser detection is applicable only for increasing the capacity on the reverse radio link and not on the forward radio link. As discussed in chapter 9, the capacity on the reverse radio link is usually less than on the forward radio link. However, for bilateral services, any improvement of the capacity on the reverse radio channel exceeding that on the forward radio channel is not justifiable. The upper bound for improving the capability on the reverse radio channel with multiuser detection is $(1 + N_I)/N_I$, where N_I is the mean interference from neighboring cells as discussed in chapter 9. For $N_I = 0.57$, which corresponds to the support of three-way soft handoffs, the upper bound equals approximately 2.75. This is the maximum gain that can be achieved; however, in practice the gain is less. An optimal multiuser detector is too complex; rather, we must accept the reduced performance of a suboptimal realization. Also, the call configuration of a base station is dynamic (i.e., new calls are set up by the base station and existing calls are handed into and out of the cell). Thus, the detector must dynamically adjust its coefficients. The wireless industry is currently exploring the utility of this approach to determine if moderate benefits can be achieved within practical bounds.

13.5 MULTIPLE BEAM ADAPTIVE ANTENNA ARRAY

An *adaptive antenna array* is defined as one that modifies its radiation pattern, frequency response, or other parameters, by means of internal feedback control while the antenna system is operating. The basic operation is described in terms of a reduction in sensitivity in certain angular position, toward a source of interference. Adaptive antennas adjust their directional beam patterns to maximize the signal-to-interference ratio at the output of the receiver. The adaptive array contains a number of antenna elements, not necessarily identical, coupled together via some form of amplitude control and phase shifting mechanism to form a single output (refer to fig. 13.4). The amplitude and phase control involves a set of complex weights. By suitable choice of weights, the array will accept a wanted signal from direction θ_1 and steer null toward interface source located at θ_k, for $k \neq 1$. The weighting mechanism can be optimized to steer

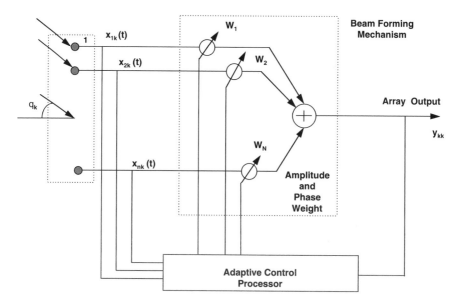

Figure 13.4 An adaptive antenna array.

beams (a radiation pattern maxima of finite width) in a specific direction, or directions. An N element array has $N - 1$ degrees-of-freedom yielding up to $N - 1$ independent pattern nulls. By controlling the weights via a feedback loop designed to maximize the signal-to-interference ratio at the array output, the system can be treated as an *adaptive spatial filter*.

The antenna elements can be arranged in various geometries, with uniform line, circular, and planar arrays being very common. The circular array geometry provides complete coverage from a central base station as beams can be steered through 360 degrees. The antenna elements are typically sited $\lambda/2$ apart, where λ represents the wavelength of the received signal. Spacing greater than $\lambda/2$ improves the spatial resolution of the array, but the formation of grating lobes (secondary maxima) can also result. These are regarded as undesirable.

Each mobile is tracked in azimuth by a narrow beam for both mobile-to-base station and base station-to-mobile transmission. The directive nature of the beams ensures that in a given system the mean interference power experienced by any one user, due to other active mobiles, would be much less than that experienced using conventional wide coverage base station antennas. Since CDMA-based cellular/PCS networks are designed to be interference limited, the adaptive antenna would considerably increase the potential user capacity.

Since the base station antenna could track any mobile or group of mobiles within its coverage area, the receiving array is capable of resolving the angular distribution of the users as they appear at the base station. The base station is then in position to form an optimal set of beams, confining the energy directed at a given mobile within a finite volume. The antenna system dynamically assigns a single narrow beam to illuminate a single mobile and broad beams to the numerous groupings of mobiles (fig. 13.5). By constraining the energy transmitted toward the mobiles, there are directions in which little or no signal is radiated. This phenomenon gives rise to the reduction in the probability of interference occurring in neighboring cells, and thereby increases the capacity of the network. A comparison made between the conventional and adaptive array antenna schemes has shown that a marked improvement in capacity of the network can be achieved. An idealized eight-beam antenna could provide a threefold increase in network capacity when compared with existing schemes such as cell splitting. The overall cost of the system is less, since fewer base stations are required for an equivalent user capacity. Also, unlike the technique of cell splitting and cell sectorization,

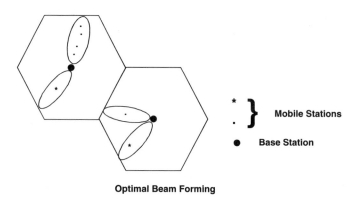

Optimal Beam Forming

Figure 13.5 Optimal beam forming.

the multiple-beam adaptive array antenna would not impair the trunking efficiency of the network.

It should be noted that this approach can be applied to improving the channel capacity on both the forward radio link and the reverse radio link, while interference cancellation, as discussed in the previous section, is limited to improving the channel capacity only on the reverse radio link.

13.6 IMPROVEMENTS OF HANDOFF ALGORITHMS

Handoffs may be initiated either by the mobile station or by the base station.[7] In the first case, the mobile station sends a *pilot strength measurement* message to the serving base station, indicating that the received signal strength of the pilot in the active set is below T_DROP or a pilot in the candidate set is above T_ADD. The base station may initiate a handoff, but it is not obligated to do so. This type of handoff is classified as a *mobile-assisted handoff*. Alternatively, measurements at the base station and at neighboring base stations, at the request of the serving base station, can initiate a handoff. This type of handoff is categorized as a *base-station-directed handoff*. Both types of handoffs are supported by current CDMA standards. A mobile-assisted handoff can be extended to a *mobile-directed handoff* in which the mobile station determines that a handoff to a selected base station will occur. The mobile-directed handoff has the potential benefit of reducing the handoff interval, but this approach has

---

7. In this discussion, the term *base station* is really the Base Transceiver System (BTS) as discussed in chapter 4.

several associated issues. First, the mobile station is determining its decision only on the forward radio link. The performance on the reverse radio link may not match that of the forward radio link at a given time. Also, the wireless infrastructure transfers handoff control to the mobile stations. This approach is vulnerable to malfunctioning mobile stations. Currently, mobile-directed handoffs are not supported by CDMA standards. A CDMA system will often combine information provided by the mobile station with information measured by the base station in determining handoff actions.

The handoff performance can often be improved by partitioning base stations into microcells and macrocells [3]. A macrocell has a large coverage area when compared with that of a microcell. In general, macrocells can overlay microcells or can abut microcells. A mobile station can be assigned to a microcell rather than a macrocell if the mobile station is traversing the cell at a slow enough velocity. If microcells and macrocells share the same frequency spectrum, then a higher transmitted power level at the microcells is necessary to compensate for interference from the macrocells. This induces a coverage hole in the macrocell in which the microcell is situated. Thus, overlaying microcells on macrocells supporting the same frequency spectrum has serious practical implications.

If the overlaid microcells use a different frequency spectrum than the macrocells, this is not an issue. This approach, however, introduces another problem. With current CDMA standards, the mobile station can monitor signals on the serving frequency. If a handoff from a microcell to a macrocell or from a macrocell to a microcell is necessary, the mobile station is not able to "assist" since it cannot measure the pilot's strength of the target base station. In this situation, the base station can either instruct neighboring base stations to measure and to report the signal strength of the mobile station, or it can use a priori information about the system configuration to determine the target base station. The first approach requires that neighboring base stations be equipped with CDMA location radios. The second approach assumes that the probability of choosing the correct target base station is high. Thus, the second approach restricts the number of system configurations that are applicable.

These issues are moot if the mobile station can measure the strength of pilots on a frequency that is different from that of the serving frequency. This capability is being discussed in CDMA standards bodies.

13.7 SUMMARY

This chapter discussed several areas that are being investigated by the industry in order to continue the evolution of CDMA technology for wireless communications. These areas include over-the-air service provisioning to streamline the activation of mobile stations, the improvement of standardized speech-coding algorithms, and the improvement of the CDMA channel capacity. The last area includes interference cancellation, multiple beam adaptive antenna arrays, and the improvement of the handoff algorithm. These approaches have technical merit, and the wireless industry is cautiously optimistic about the corresponding benefits in the future.

However, before we conclude this chapter, it is important that we do not lose track of the reasons that we adopt new technologies. The reason is not to challenge the engineering community. The real reasons are to solve problems and satisfy the needs of the customer and to provide better service at the same or lower cost. As an example, capacity improvement is not a goal itself but rather a means to a goal (i.e., a reduced price to the customer). The future of the wireless industry is dependent upon holding true to this philosophy.

13.8 REFERENCES

1. Duel-Hallen, A., Holtzman, J., and Zvonar, Z., "Multiuser Detection for CDMA Systems," *IEEE Personal Communications*, April 1993.
2. Naguib, A. F., Paulraj, A., and Kailath, T., "Capacity Improvement with Base–Station Antenna Arrays in Cellular CDMA", *IEEE Trans., Veh., Technol.*, *VT-43*, 1994, pp. 691–698.
3. Pollini, G. P., "Trends in Handover Design," *IEEE Communications Magazine*, March 1996.
4. Swales, S. C., Beach, M. A., Edwards, D. J., and McGeehan, J. P., "The Performance Enhancement of Multibeam Adaptive Base–Station Antennas for Cellular Land Mobile Radio Systems," *IEEE Trans., Veh., Technol.*, *VT-39*, 1990, pp. 56–66.
5. TIA PN-3646, "Enhanced Variable Rate Codec (EVRC), Service Option 3—Draft," December 1995.
6. TIA PN-3569, "Over-the-Air Service Provisioning of Mobile Stations in WBSS Systems—Ballot Text," September 1, 1995.
7. TIA TR45.5, Contribution TR45.5.1.1/95.07.17.03, "EVRC Host and Listening Laboratories—Draft Interim Report," July 1995.

Traffic Tables

This appendix[1] provides traffic tables for a variety of blocking probabilities and channels. The blocked calls cleared (Erlang B) call model is used. With the Erlang B model, we assume that when traffic arrives in the system, it either is served, with probability from the table, or is lost to the system. A customer attempting to place a call will therefore either see a call completion or be blocked and abandon the call. This assumption is acceptable for low blocking probabilities. In some cases, the call will be placed again after a short period of time. If too many calls reappear in the system after a short delay, the Erlang B model will no longer hold.

In Tables A.8 and A.9, where the number of channels is high (greater than 250 channels), linear interpolation between two table values is possible. We provide the deltas for one additional channel to assist in the interpolation.

---

1. The data in the tables was supplied by V. H. MacDonald.

Table A.1 Offered Loads (in Erlangs) for Various Blocking Objectives: According to the Erlang B Model—
System Capacity for 1–20 Channels

Trunks	0.01	0.015	0.02	0.03	0.05	0.07	0.1	0.2	0.5
P(B)=									
1	0.010	0.015	0.020	0.031	0.053	0.075	0.111	0.250	1.000
2	0.153	0.190	0.223	0.282	0.381	0.471	0.595	1.000	2.732
3	0.455	0.536	0.603	0.715	0.899	1.057	1.271	1.930	4.591
4	0.870	0.992	1.092	1.259	1.526	1.748	2.045	2.944	6.501
5	1.361	1.524	1.657	1.877	2.219	2.504	2.881	4.010	8.437
6	1.913	2.114	2.277	2.544	2.961	3.305	3.758	5.108	10.389
7	2.503	2.743	2.936	3.250	3.738	4.139	4.666	6.229	12.351
8	3.129	3.405	3.627	3.987	4.543	4.999	5.597	7.369	14.318
9	3.783	4.095	4.345	4.748	5.370	5.879	6.546	8.521	16.293
10	4.462	4.808	5.084	5.529	6.216	6.776	7.511	9.684	18.271
11	5.160	5.539	5.842	6.328	7.076	7.687	8.487	10.857	20.253
12	5.876	6.287	6.615	7.141	7.950	8.610	9.477	12.036	22.237
13	6.607	7.049	7.402	7.967	8.835	9.543	10.472	13.222	24.223
14	7.352	7.824	8.200	8.803	9.730	10.485	11.475	14.412	26.211
15	8.108	8.610	9.010	9.650	10.633	11.437	12.485	15.608	28.200
16	8.875	9.406	9.828	10.505	11.544	12.393	13.501	16.807	30.190
17	9.652	10.211	10.656	11.368	12.465	13.355	14.523	18.010	32.181
18	10.450	11.024	11.491	12.245	13.389	14.323	15.549	19.215	34.173
19	11.241	11.854	12.341	13.120	14.318	15.296	16.580	20.424	36.166
20	12.041	12.680	13.188	14.002	15.252	16.273	17.614	21.635	38.159

Table A.2 Offered Loads (In Erlangs) for Various Blocking Objectives: According to the Erlang B Model—System Capacity for 20–39 Channels

Trunks	0.005	0.01	0.015	0.02	0.03	0.05	0.07	0.1
P(B)=								
20	11.092	12.041	12.680	13.188	14.002	15.252	16.273	17.614
21	11.860	12.848	13.514	14.042	14.890	16.191	17.255	18.652
22	12.635	13.660	14.352	14.902	15.782	17.134	18.240	19.693
23	13.429	14.479	15.196	15.766	16.679	18.082	19.229	20.737
24	14.214	15.303	16.046	16.636	17.581	19.033	20.221	21.784
25	15.007	16.132	16.900	17.509	18.486	19.987	21.216	22.834
26	15.804	16.966	17.758	18.387	19.395	20.945	22.214	23.885
27	16.607	17.804	18.621	19.269	20.308	21.905	23.214	24.939
28	17.414	18.646	19.487	20.154	21.224	22.869	24.217	25.995
29	18.226	19.493	20.357	21.043	22.143	23.835	25.222	27.053
30	19.041	20.343	21.230	21.935	23.065	24.803	26.229	28.113
31	19.861	21.196	22.107	22.830	23.989	25.774	27.239	29.174
32	20.685	22.053	22.987	23.728	24.917	26.747	28.250	30.237
33	21.512	22.913	23.869	24.629	25.846	27.722	29.263	31.302
34	22.342	23.776	24.755	25.532	26.778	28.699	30.277	32.367
35	23.175	24.642	25.643	26.438	27.712	29.678	31.294	33.435
36	24.012	25.511	26.534	27.346	28.649	30.658	32.312	34.503
37	24.852	26.382	27.427	28.256	29.587	31.641	33.331	35.572
38	25.694	27.256	28.322	29.168	30.527	32.624	34.351	36.643
39	26.539	28.132	29.219	30.083	31.469	33.610	35.373	37.715

Table A.3 Offered Loads (In Erlangs) for Various Blocking Objectives: According to the Erlang B Model— System Capacity for 40–60 Channels

Trunks	0.005	0.01	0.015	0.02	0.03	0.05	0.07	0.1
P(B)=								
40	27.387	29.011	30.119	30.999	32.413	34.597	36.397	38.788
41	28.237	29.891	31.021	31.918	33.359	35.585	37.421	39.861
42	29.089	30.774	31.924	32.838	34.306	36.575	38.447	40.936
43	29.944	31.659	32.830	33.760	35.255	37.565	39.473	42.012
44	30.801	32.546	33.737	34.683	36.205	38.558	40.501	43.088
45	31.660	33.435	34.646	35.609	37.156	39.551	41.530	44.165
46	32.521	34.325	35.556	36.535	38.109	40.545	42.559	45.243
47	33.385	35.217	36.468	37.463	39.063	41.541	43.590	46.322
48	34.250	36.111	37.382	38.393	40.019	42.537	44.621	47.401
49	35.116	37.007	38.297	39.324	40.976	43.535	45.654	48.481
50	35.985	37.904	39.214	40.257	41.934	44.534	46.687	49.562
51	36.856	38.802	40.132	41.190	42.893	45.533	47.721	50.644
52	37.728	39.702	41.052	42.125	43.853	46.533	48.756	51.726
53	38.601	40.604	41.972	43.061	44.814	47.535	49.791	52.808
54	39.477	41.507	42.894	43.999	45.777	48.537	50.827	53.891
55	40.354	42.411	43.817	44.937	46.740	49.540	51.864	54.975
56	41.232	43.317	44.742	45.877	47.704	50.544	52.902	56.059
57	42.112	44.224	45.667	46.817	48.669	51.548	53.940	57.144
58	42.993	45.132	46.594	47.759	49.636	52.553	54.979	58.229
59	43.875	46.041	47.522	48.701	50.603	53.559	56.018	59.315
60	44.759	46.951	48.451	49.645	51.570	54.566	57.058	60.401

Table A.4 Offered Loads (In Erlangs) for Various Blocking Objectives: According to the Erlang B Model—System Capacity from 61–80 Channels

Trunks	0.005	0.01	0.015	0.02	0.03	0.05	0.07	0.1
P(B)=								
61	45.644	47.863	49.381	50.590	52.539	55.573	58.099	61.488
62	46.531	48.776	50.311	51.535	53.509	56.581	59.140	62.575
63	47.418	49.689	51.243	52.482	54.479	57.590	60.181	63.663
64	48.307	50.604	52.176	53.429	55.450	58.599	61.224	64.750
65	49.197	51.520	53.110	54.377	56.422	59.609	62.266	65.839
66	50.088	52.437	54.044	55.326	57.395	60.620	63.309	66.927
67	50.980	53.355	54.980	56.276	58.368	61.631	64.353	68.016
68	51.874	54.273	55.916	57.226	59.342	62.642	65.397	69.106
69	52.768	55.193	56.853	58.178	60.316	63.654	66.442	70.196
70	53.663	56.113	57.791	59.130	61.292	64.667	67.487	71.286
71	54.560	57.035	58.730	60.083	62.268	65.680	68.532	72.376
72	55.457	57.957	59.670	61.036	63.244	66.694	69.578	73.467
73	56.356	58.880	60.610	61.991	64.222	67.708	70.624	74.558
74	57.255	59.804	61.551	62.945	65.199	68.723	71.671	75.649
75	58.155	60.729	62.493	63.901	66.178	69.738	72.718	76.741
76	59.056	61.654	63.435	64.857	67.157	70.753	73.765	77.833
77	59.958	62.581	64.379	65.814	68.136	71.769	74.813	78.925
78	60.861	63.508	65.322	66.772	69.116	72.786	75.861	80.018
79	61.765	64.435	66.267	67.730	70.097	73.803	76.909	81.110
80	62.669	65.364	67.212	68.689	71.078	74.820	77.958	82.203

Table A.5 Offered Loads (In Erlangs) for Various Blocking Objectives: According to the Erlang B Model—System Capacity for 81–100 Channels

Trunks	0.005	0.01	0.015	0.02	0.03	0.05	0.07	0.1
P(B)=								
81	63.574	66.293	68.158	69.648	72.059	75.838	79.007	83.297
82	64.481	67.223	69.104	70.608	73.042	76.856	80.057	84.390
83	65.387	68.153	70.051	71.568	74.024	77.874	81.107	85.484
84	66.295	69.085	70.999	72.529	75.007	78.893	82.157	86.578
85	67.204	70.016	71.947	73.491	75.991	79.912	83.207	87.672
86	68.113	70.949	72.896	74.453	76.975	80.932	84.258	88.767
87	69.023	71.882	73.846	75.416	77.959	81.952	85.309	89.861
88	69.933	72.816	74.796	76.379	78.944	82.972	86.360	90.956
89	70.844	73.750	75.746	77.342	79.929	83.993	87.411	92.051
90	71.756	74.685	76.697	78.306	80.915	85.014	88.463	93.146
91	72.669	75.621	77.649	79.271	81.901	86.035	89.515	94.242
92	73.582	76.557	78.601	80.236	82.888	87.057	90.568	95.338
93	74.496	77.493	79.553	81.202	83.875	88.079	91.620	96.434
94	75.411	78.431	80.506	82.167	84.862	89.101	92.673	97.530
95	76.326	79.368	81.460	83.134	85.850	90.123	93.726	98.626
96	77.242	80.307	82.414	84.101	86.838	91.146	94.779	99.722
97	78.158	81.245	83.368	85.068	87.827	92.169	95.833	100.819
98	79.075	82.185	84.323	86.036	88.815	93.193	96.887	101.916
99	79.993	83.125	85.279	87.004	89.805	94.217	97.941	103.013
100	80.911	84.065	86.235	87.972	90.794	95.240	98.995	104.110

Table A.6 Offered Loads (In Erlangs) for Various Blocking Objectives: According to the Erlang B Model—System Capacity for 105–200 Channels

Trunks	0.005	0.01	0.015	0.02	0.03	0.05	0.07	0.1
P(B)=								
105	85.518	88.822	91.030	92.823	95.747	100.371	104.270	109.598
110	90.147	93.506	95.827	97.687	100.713	105.496	109.550	115.090
115	94.768	98.238	100.631	102.552	105.680	110.632	114.833	120.585
120	99.402	102.977	105.444	107.426	110.655	115.772	120.121	126.083
125	104.047	107.725	110.265	112.307	115.636	120.918	125.413	131.583
130	108.702	112.482	115.094	117.195	120.622	126.068	130.708	137.087
135	113.366	117.247	119.930	122.089	125.615	131.222	136.007	142.593
140	118.039	122.019	124.773	126.990	130.612	136.380	141.309	148.101
145	122.720	126.798	129.622	131.896	135.614	141.542	146.613	153.611
150	127.410	131.584	134.477	136.807	140.621	146.707	151.920	159.122
155	132.106	136.377	139.337	141.724	145.632	151.875	157.230	164.636
160	136.810	141.175	144.203	146.645	150.647	157.047	162.542	170.152
165	141.520	145.979	149.074	151.571	155.665	162.221	167.856	175.668
170	146.237	150.788	153.949	156.501	160.688	167.398	173.173	181.187
175	150.959	155.602	158.829	161.435	165.713	172.577	178.491	186.706
180	155.687	160.422	163.713	166.373	170.742	177.759	183.811	192.227
185	160.421	165.246	168.602	171.315	175.774	182.943	189.133	197.750
190	165.160	170.074	173.494	176.260	180.809	188.129	194.456	203.273
195	169.905	174.906	178.390	181.209	185.847	193.318	199.781	208.797
200	174.653	179.743	183.289	186.161	190.887	198.508	205.108	214.323

Table A.7 Offered Loads (in Erlangs) for Various Blocking Objectives: According to the Erlang B Model—
System Capacity for 205–245 Channels

Trunks	0.005	0.01	0.015	0.02	0.03	0.05	0.07	0.1
P(B)=								
205	179.407	184.584	188.192	191.116	195.930	203.700	210.436	219.849
210	184.165	189.428	193.099	196.073	200.976	208.894	215.765	225.376
215	188.927	194.276	198.008	201.034	206.023	214.089	221.096	230.904
220	193.694	199.127	202.920	205.997	211.073	219.287	226.427	236.433
225	198.464	203.981	207.836	210.963	216.125	224.485	231.760	241.963
230	203.238	208.839	212.754	215.932	221.180	229.686	237.094	247.494
235	208.016	213.700	217.675	220.902	226.236	234.887	242.430	253.025
240	212.797	218.564	222.598	225.876	231.294	240.090	247.766	258.557
245	217.582	223.430	227.524	230.851	236.354	245.295	253.103	264.089

Table A.8 Offered Loads (In Erlangs) for Various Blocking Objectives: According to the Erlang B Model—System Capacity for 250–600 Channels

Trunks	0.005	0.01	0.015	0.02	0.03	0.05	0.07	0.1
P(B)=								
250	222.370	228.300	232.452	235.828	241.415	250.500	258.441	269.622
Δ	0.961	0.977	0.988	0.998	1.015	1.042	1.069	1.107
300	270.410	277.144	281.853	285.707	292.142	302.617	311.866	324.961
Δ	0.966	0.980	0.991	1.001	1.017	1.044	1.070	1.108
350	318.698	326.155	331.424	335.738	342.995	354.836	365.359	380.384
Δ	0.969	0.984	0.994	1.005	1.018	1.045	1.071	1.109
400	367.163	375.334	381.128	385.963	393.895	407.096	418.890	435.813
Δ	0.972	0.989	0.998	1.004	1.020	1.046	1.071	1.109
450	415.779	424.774	431.022	436.178	444.877	459.408	472.456	491.263
Δ	0.975	0.987	0.997	1.006	1.021	1.047	1.072	1.109
500	464.518	474.130	480.890	486.480	495.919	511.759	526.049	546.730
Δ	0.977	0.989	0.999	1.007	1.022	1.048	1.072	1.110
550	513.361	523.600	530.843	536.846	547.012	564.142	579.663	602.208
Δ	0.979	0.991	1.000	1.008	1.023	1.048	1.073	1.110
600	562.292	573.142	580.859	587.267	598.145	616.552	633.295	657.697

Table A.9 Offered Loads (In Erlangs) for Various Blocking Objectives: According to the Erlang B Model—System Capacity for 600–1050 Channels

Trunks	0.005	0.01	0.015	0.02	0.03	0.05	0.07	0.1
P(B)=								
600	562.292	573.142	580.859	587.267	598.145	616.552	633.295	657.697
Δ	0.983	0.992	1.001	1.009	1.023	1.049	1.073	1.110
650	611.418	622.748	630.927	637.732	649.313	668.982	686.941	713.193
Δ	0.981	0.993	1.002	1.010	1.024	1.049	1.073	1.110
700	660.462	672.410	681.042	688.238	700.511	721.432	740.598	768.697
Δ	0.982	0.994	1.003	1.011	1.024	1.049	1.073	1.110
750	709.586	722.119	731.196	738.777	751.735	773.896	794.266	824.206
Δ	0.984	0.995	1.004	1.011	1.025	1.050	1.074	1.110
800	758.762	771.872	781.386	789.346	802.981	826.375	847.943	879.719
Δ	0.985	0.996	1.004	1.012	1.025	1.050	1.074	1.110
850	807.987	821.662	831.608	839.942	854.247	878.865	901.627	935.236
Δ	0.985	0.996	1.005	1.012	1.026	1.050	1.074	1.110
900	857.256	871.487	881.857	890.561	905.530	931.365	955.317	990.757
Δ	0.986	0.997	1.005	1.013	1.026	1.050	1.074	1.110
950	906.565	921.343	932.132	941.202	956.829	983.875	1009.013	1046.281
Δ	0.987	0.998	1.006	1.013	1.026	1.050	1.074	1.111
1000	955.910	971.226	982.430	991.862	1008.142	1036.393	1062.715	1101.808
Δ	0.988	0.998	1.006	1.014	1.027	1.050	1.074	1.111
1050	1005.289	1021.136	1032.748	1042.539	1059.468	1088.918	1116.420	1157.337

List of Abbreviations
and Acronyms

A

ABR	Answer Bid Ratio
AbS	Analysis by Synthesis
AC	Authentication Center
ACM	Address Complete Message
ADPCM	Adaptive Differential Pulse Code Modulation
AMPS	Advanced Mobile Phone System
APC	Adaptive Predictive Coding
APC	Automatic Power Control
ARLAN	Advanced Radio LAN
ARPA	Advanced Research Project Agency
ARQ	Automatic Repeat Request
ATIS	Alliance for Telecommunications Industry Solutions
ASR	Answer Seizure Ratio
AWGN	Additive White Gaussian Noise

B

BER	Bit-Error Rate
BHCA	Busy Hour Call Attempts
BML	Business Management Layer
BPSK	Binary Phase-Shift Keying
BS	Base Station
BSAP	Base Station Application Part
BSC	Base Station Controller
BSMAP	Base Station Management Application Part
BTS	Base Transceiver System

C

CDCP	Call Detail Collection Point
CDGP	Call Detail Generation Point
CDIS	Call Detail Information Source
CDMA	Code-Division Multiple Access
CDRP	Call Detail Rating Point
CELP	Code-Excited Linear Prediction
CGI	Call Global Identification
CI	Call Identity
CM	Channel Modem
CNIP	Calling Number Identification Presentation
CNIR	Calling Number Identification Restriction
CRC	Cyclic Redundancy Check
CSMA/CA	Carrier-Sensed Multiple Access with Collision Avoidance

D

DCE	Data Communication Equipment
DCN	Data Communications Network
DMH	Data Message Handler
DN	Directory Number
DSSS	Direct Sequence Spread Spectrum
DTAP	Direct Transfer Application Part
DTE	Data Terminal Equipment
DTMF	Dual-Tone Multifrequency

E

EIA	Electronic Industry Association
EIR	Equipment Identity Register
EIRP	Effective Isotropic Radiated Power
EML	Element Management Layer
ERP	Effective Radiated Power
ESN	Electronic Serial Number
EVRC	Enhanced Variable Rate Codec

F

FAF	Floor Attenuation Factor
FCC	Federal Communications Commission
FCS	Frame Check Sequence
FD	Full Duplex
FDD	Frequency-Division Duplex
FDMA	Frequency-Division Multiple Access

FEC	Forward Error Correction
FHSS	Frequency-Hopping Spread Spectrum
FM	Frequency Modulation

G

GBN	Go-Back-N
GFSK	Gaussian Frequency Shift Keying
GHz	gigahertz
GMSK	Gaussian Minimum Shift Keying
GOS	Grade of Service
GPS	Global Positioning System
GSM	Global System of Mobile Communications

H

HAAT	Height Above Average Terrain
HD	Half Duplex
HIPERLAN	High-Performance Radio Local Area Network
HLR	Home Location Register

I

IAM	Initial Address Message
ICMP	Internet Control Message Protocol
IDs	Identifications
IEEE	Institute of Electrical and Electronic Engineers
IF	Intermediate Frequency
IMSI	International Mobile Station Identification
IN	Intelligent Network
IP	Internet Protocol
IPCP	Internet Protocol Control Protocol
ISDN	Integrated Services Digital Network
ISM	Industrial Scientific Medical
ISO	International Standard Organization
IWF	InterWorking Function

K

kbps	kilobit per second
kHz	kilohertz

L

LAC	Local Area Code
LAN	Local Area Network

LAC	Linear Amplifier Circuit
LBT	Listen-Before-Talk
LCP	Link Control Protocol
LOS	Line of Sight
LPC	Linear Predictive Coding
LSP	Line Spectral Pairs

M

MAC	Media Access
MAN	Metropolitan Area Network
MAP	Mobile Application Part
MC	Message Center
MCC	Mobile Country Code
Mcps	million chips per second
MF	Multifrequency
MF	Mediation Function
MHz	megahertz
MIN	Mobile Identification Number
MM	Mobility Management
MMAP	Mobility Management Application Part
MNC	Mobile Network Code
MNRP	Mobile Network Registration Protocol
MOS	Mean Opinion Score
MS	Mobile Station
MSC	Mobile Switching Center
MTP	Message Transfer Part

N

NAM	Number Assignment Module
NE	Network Element
NEF	Network Element Function
NEL	Network Element Layer
NID	Network Identification
NML	Network Management Layer

O

OA&M	Operation, Administration, & Maintenance
OAM&P	Operation, Administration, Maintenance, & Provisioning
OQPSK	Offset Quadrature Phase-Shift Keying
OS	Operations System
OSF	Operations Systems Function

OSI	Open System Interconnect
OTASP	Over-The-Air Service Provisioning

P

PBX	Private Branch Exchange
PCM	Pulse Code Modulation
PCMCIA	Personal Computer Memory Card International Association
PCN	Personal Communications Network
PCS	Personal Communications Services
PCSAP	PCS Application Part
PDA	Personal Digital Assistants
PIN	Personal Identification Number
PLMN	Public Land Mobile Network
PMC	Personal Mobility Controller
PN	Pseudorandom Noise
PPDN	Public Packet Data Network
PPP	Point-to-Point Protocol
PSPDN	Public Switched Packet Data Network
PSTN	Public Switched Telephone Network

Q

QAF	Q Adapter Function
QCELP	QUALCOMM Code-Excited Linear Prediction
QoS	Quality of Service
QPSK	Quadrature Phase-Shift Keying

R

RASC	Radio Access System Controller
RELP	Residual-Excited Linear Prediction
RLP	Radio Link Protocol
RPC	Radio Port Controller

S

SCCP	Signaling Connection Control Part
SCI	Synchronized Capsule Indicator
SCM	Station Class Mark
SID	System Identification
SML	Service Management Layer
SMRS	Specialized Mobile Radio Services
SMS	Short Message Service

SNDCF	SubNetwork Dependent Convergence Function
SOF	Start Of Frame
SOM	Start Of Message
SOS	Start Of Slot
SRP	Selective Repeat Request
SS	Spread Spectrum
SSD	Shared Secret Data
SS7	Signaling System 7

T

TA	Terminal Adapter
TCP	Transmission Control Protocol
TDMA	Time-Division Multiple Access
TE	Terminal Equipment
TH	Time Hopped
TIA	Telecommunications Industry Association
TMA	Telesystems Microcellular Architecture
TMC	Terminal Mobility Controller
TMN	Telecommunications Management Network
TMSI	Temporary Mobile Station Identification
T-R	Transmitter Receiver

U

UCB	University of California at Berkeley
UCLA	University of California at Los Angeles

V

VLR	Visitor Location Register

W

WAN	Wide Area Network
W-CDMA	Wideband CDMA
WIN	Wireless Intelligent Network
WLAN	Wireless Local Area Network
WORM	Window-control Operation based on Reception Memory
WPBX	Wireless Private Branch Exchange
WS	Work Station
WSF	Work Station Function

X

XC	Transcoder

Index

A

Access channel, 57, 103–105, 128–129
  W-CDMA, 136–138
Access methods, 246–255
  ALOHA, pure, 248
  ALOHA, slotted, 248–249
  AT&T WaveLAN, 237–240
  carrier-sensed multiple access, 249–251
Access parameters message, 56, 112, 125
Acknowledgment procedures, 51–52
Adaptive antenna array, 322–324
Adaptive differential pulse code modulation, 81, 166, 167
Adaptive predictive coding, 168
Additive white Guassian noise, 18
Address complete message, 145, 147
Advanced mobile phone system (AMPS), 50, 313
Advanced Research Project Agency, 245
A-interface, 74, 76–86
A-key, 325
Alliance for Telecommunications Industry Association, 93

Analysis by synthesis, 171
Antenna gain, 178
Asynchronous data, 256–261
Authentication center, 74, 325
Authentication challenge message, 127
Authentication challenge response message, 130
Automatic repeat request, 251
Autonomous registration, 64–68
  distance-based, 67
  power-down, 66
  power-up, 66
  timer-based, 66
  zone-based, 67

B

Bandpass filtering, 118
Base station, 72–73, 79
Base station application part, 78
Base station-associated handoff, 157
Base station controller, 72–73, 79
Base station-directed handoff, 334
Base station management application, 78
Base transceiver system, 73, 79

353

Binary phase-shift keying,
    coherent, 15–16
Bit-error rate
    binary phase-shift keying, 16
    quadrature phase-shift keying,
    18
Bit repetition, 115
Block interweaving, 99, 115
Busy hour call attempt, 209

C

Calling number identification
    presentation, 89
Calling number identification
    restriction, 89–90
Call model, IS-95A, 53–59
    mobile station idle state, 55
    mobile station initialization
    state, 54
    system access state, 57
    traffic channel state, 57
Capacity of forward link, 229–230
Capacity of reverse link, 218–225
CDMA. See Code-division multiple
    access
CDMA channel list message, 56
CDMA to analog handoff, 50
Cell global identification, 81
Cell identity, 81
Channel assignment message, 126
Channel modem, 225
Characteristic polynomial, 45
Chip, 30
    rate, 40, 42, 44
    sequence, 28
Clearing, call procedure, 147–149
Code-division multiple access
    (CDMA)
    performance, 19–23
    soft limit, 217

Code-excited linear prediction,
    166, 171–174, 327
Convolutional coding, 41, 98, 103,
    113–114
Cookbook, index, 174
Customer service center, 324
Cyclic redundancy check, 122–
    123, 126, 128, 133

D

Data link connection identifier, 78
Data link layer, 51, 95
Data message handler, 74, 288–
    292
Data terminal equipment, 256
Diffie-Hellman key exchange
    procedure, 325–326
Digital technologies,
    different approaches, 4–6
    needs, 3
Direct sequence spread spectrum,
    9
    performance, 15–19
    WLAN, 236
Directory number, 45
Direct transfer application part,
    78
Diversity reception, 80
Downlink, IS-95A, 41
Dual mode digital systems, 315–
    316
Dual-tone multifrequency, 75, 310
Dynamic channel assignment, 200

E

Effective interference power, 19
Effective isotropic radiated power,
    110–111, 112
Effective radiated power, 191, 195
EIA-553 (IS-91) standard, 313